THIRSTY NATION

THIRSTY NATION

PRIORITIES FOR INDIA'S WATER SECTOR

JOSEPH P. QUINLAN
SUMANTRA SEN
KIRAN NANDA

RANDOM HOUSE INDIA

Published by Random House India in 2014
1

Random House Publishers India Private Limited
Windsor IT Park, 7th Floor, Tower-B
A-1, Sector-125, Noida-201301 (UP)

Random House Group Limited
20 Vauxhall Bridge Road
London SW1V 2SA
United Kingdom

978 81 8400 354 3

Typeset in Adobe Caslon Pro by Saanvi Graphics, Noida

Printed and bound in India by Replika Press Private Limited

Contents

Why This Book?

Cherian Thomas, CEO, IDFC Foundation

'Ensuring access to clean and potable water is a continuing challenge for a large part of the developing world, including India, given the limited availability of fresh water sources. This challenge is exacerbated by unsustainable agricultural practices requiring intensive water use and the pollution of fresh water sources due to the discharge of untreated industrial and household effluents. The disuse and neglect of traditional localized sources like tanks and ponds and equally their conversion, especially in urban areas, to commercial real estate have further compounded the problem. It is expected that the growth in populations and their higher aspirations following economic growth would result in even more intensive water use over the years, making it an increasingly scarce commodity and, as some have gloomily predicted, a cause for strife and conflict, both local and global.

Water management—riparian rights and sharing of water, priority of usage, user charges, water harvesting, recharging, recycling and treatment throughout its physical and economic value chain, therefore, needs increasing attention, going forward. Credible research in these areas for setting out successful and sustainable management practices could help more optimal utilization of this resource. Equally important would be its role in shaping policies, projects and programmes to ensure conservation, protection and sustainable management of this resource. Informed and timely action is clearly the need of the hour.'

Namita Vikas, Senior President and Country Head-Responsible Banking, YES Bank Limited

'In a country that believes in:

Not the donation of land, Not the donation of gold,

Not the donation of cows, Not even the donation of food,

Donation of water is the greatest of all donations in this world, the throes of evolving into a developing nation is making India water-scarce, where an average villager walks about 2.5 kms for clean drinking water. Despite the abundance of resources, Indian states stand at economic crossroads of inability to share water equitably. This along with political uncertainties puts development in the backseat. It is not only the responsibility of government alone to address water scarcity, but also private sector involvement is the need of the hour. While the private sector is capable of dealing with water use and treatment aspects, government needs to effectively handle the core issue and work towards replenishing water tables. With the demand for water only increasing, financial institutions, being conduits of capital, need to look at innovative services and solutions to water-financing like promoting newer cost-effective technologies and providing preferential debt solutions to water management companies.

"Thirsty Nation" comes at a very strategic point in time, where India's water policy is taking shape to be comprehensive, responding to the exponential increase in demand. This book serves as a propeller of water-financing, quoting global expertise and suggesting Indianised variants. It also provides key insights to the readers to adopt and adapt.'

Dr Abhay Pethe, Professor of Urban Economics and Regional Development, VibhutiShukla Chair

'For some time to come, water will engage the Indian economy, polity and society as a major area of concern. Applying technology and innovation to both demand and supply sides of this sector and creation of incentivizing policy framework is of essence. This encompasses production/ extraction, capture/ storage and water-usage. Water embodies consumption or 'development entitlement' aspect as well as productive resources or 'investment' aspect making it a complex good. Getting water-pricing right is at the heart of the matter. This alone will enable bankable water projects to emerge which will attract private investment in this most important sector.'

Niranjan Hiranandani, President, IMC

'Water is one of the vital inputs for economic development of a modern society. It signifies two aspects—first the water management and second the ecological aspect. Both these aspects must be coupled with the quality and quantity of water to make the impact powerful. Water is not only a problem at city or country level but also at global level. The entire lifecycle of water sustainability interventions need to be studied and followed up, where Business Membership organizations (BMOs) like Chambers can play an important role both as enablers and catalysts.

As there is a great need to expand and improve the water supply network and to increase the efficiency of existing water use, a holistic approach to projects management and its execution has become mandatory. This approach must be right from the inception stage to water-sourcing, transportation and distribution equitably and inclusively, metering, revenue collection, water treatment, recycling and till last stage of water efficient usage by people.

Chambers of Commerce and Industry can play a responsible role encompassing information gathering that can feed-in at every stage of lifecycle of interventions; collating information about technicalities, economics and resource persons and making it available to all stakeholders including policy-making authorities, resulting in its fruitful usage for the benefit of the nation.

There are ladders of intervention which any organization as mature corporate citizen needs to engage in. Indian Merchants' Chamber is one such mega organization. Companies or company organizations ought to be engaging into dialogues or sponsoring studies, which lead to submissions that strengthen sustainable water decisions. The results may yield results that are unfavourable to industries in the short term by requiring them to pay more than current charges. However, these kinds of engagements have a potential to benefit industries over a period in myriad sustainable ways.

The crux of water sustainability is pricing of water and has become a pressing issue of the time in India. Any study on "what would be the genuine water pricing?" could lead to sustainable tariff setting models. The role of Chamber could be to develop a kind of water pricing model which is based on transparent subsidies and is also equitable. Rather than adopting a model that is developed elsewhere, at IMC we envisage developing a Viable-India-Model.

Chambers should be providing facilitating platform for brainstorming sessions amongst all stakeholders towards offering innovative investment solutions for water, taking such initiatives that contribute to attaining national water sustainability and facilitate 24×7 affordable quality water supply to the residents.'

Dr Ram Babu, CEO- General Carbon Advisory Services

'The importance of responsible behaviour in drawing, using and discharging water cannot be over emphasized. The entities that draw, use and discharge water in large quantities (industries and other institutions) should adopt water stewardship processes which include:

- Understanding the impacts of water withdrawal from fresh sources (direct or indirect) and minimizing such impacts
- Improving water use efficiency by various means including recycle and reuse
- Being responsible in discharging the quantum and quality of water to environment (land/water bodies) and its impacts
- And finally, compensating the unavoidable impacts by supporting recharge, rejuvenate, reuse and recycle outside the sphere of control

To promote water resource stewardship amongst all major managers and users of water resources, synergy amongst multiple water resource and environment management institutions and compatibility of policies and legislations are essential. Water foot printing (mapping "as-is" of entities water resource stewardship) and variety of market based instruments were proven very useful in industrial and commercial sectors.'

Ramani Iyer, Forbes Marshall and Mentor for Water

Until we create confidence for any urban dweller in India, that he can safely drink water from his domestic tap, 24×7, I would consider our democracy to be incomplete. We must also eliminate the scourge of women walking for water in all parts of India. Till that is achieved all talk of social equity is empty rhetoric.

Sameer Unhale, Officer on Special Duty, JNNURM, State Level Nodal Agency, MMRDA, Maharashtra

A project-oriented approach does not always take an integrated and holistic view of the urban sector. This approach fragments the viewing of the water sector. Water management is 'everyone's responsibility' and requires holistic perspective for arriving at meaningful solutions.

Similarly, the citizens' perceptions on water services, does not get factored in the conceptualization of the project. It is said, customer is king. But, the citizen is still an applicant in the economic history, encompassing a period of over six decades, since the start of the first five year plan in 1951. What is conspicuously absent is an interface of management with final users of water.

It also is important to study the following parameters and apply, i.e., whether it's possible and feasible to have three types of water supply for the households.

- Pipeline for potable water
- Pipeline for water being used for bathing purposes, but quality need not be same as that of potable water
- Pipeline for water being used for flushing purposes, but quality need not be same as that of potable water and water being used for bathing

Water used from second and third pipelines can be sent back to the source of supply by a common pipeline, so that it can be reused. This can provide a cost effective solution to water woes of the country.

Suresh Prabhu, Former Union Minster, Chairperson, Council for Energy, Environment and Water (CEEW), and Chairman, National River Linking Project

Water is not only a problem at city or country level, but also at the global level. Water signifies two aspects—one a human-centered view, i.e., water management, and second, the ecological aspect—a key element of nature—which is akin only to wind. In my view, both aspects—water management as well as ecological aspects of water, also coupled with qualitative aspects of water, in addition to enhanced quantity, all need to be considered together, to make the impact useful.

Climate change has a profound effect, especially on irrigated agriculture, industry and people in the form of floods, droughts, and rise in temperature. Combating and adapting to climate change in the form of adaptation and mitigation strategies need to be worked out.

As we can observe, 'Water' related issues cut across various ministries and departments without an effective coordination and creation of synergy. As a result, the implementation of various programs remains inefficient and causes undue delays. Creation of synergy amongst ministries and others, on water, would also help in mainstreaming various sectoral programs.

The uses of water by industry and urban areas needs to be made much more efficient. This requires reducing dependence on fresh water, especially ground water and ensuring safe disposal of waste. A massive push is needed to attract private investment both domestic and foreign in all areas of water infrastructure, both large projects, and for drinking water supply. Investments will flow in the water sector only if the sector becomes viable and efficient.

As India is already a water stressed country, there is great need to expand and improve the water supply network, increase the efficiency of existing water use and reduce demand by shifting the country and its cities towards less water intensive activities, through management and technological changes.

Limited water supplies and variety of challenges, including competing agricultural, environmental, industrial, and municipal interests present a formidable state of India's water management.

A crisp and focused research is required on the current water woes of the country, and practical implementable measures are needed towards sustainable practices and strategies for efficient water resources management in the country.

A low-water economy is the need of the hour. If the twentieth century was dominated by anxiety over oil, the twenty-first century will be consumed by concerns over the declining quantity and quality of our water resources.

Though water is a local issue, but its management needs to have a holistic perspective. In my view, the 12th FYP's national water strategy, rightfully places water at the heart of development planning. The thrust on water policy and management needs to be implemented in the right spirit, to result in true inclusive economic growth.

Shawahiq Siddiqui, Environmental Lawyer, and Founder Partner, Indian Environment Law

The Planning Commission's approach towards the twelfth five year plan on water is endorsed in the working group papers submitted to it on i) the National Water Framework Legislation, ii) the model Ground Water Bill, and iii) a framework for water governance. These documents provide a clear indication that water is increasingly being looked as a national resource, having economic value, and integral to the development and well-being of the country, by the planners and policy makers.

Legally speaking, water is a state subject under the Constitution. It means that power to legislate on water rests with the state governments. Further, water related provisions are spread in different state agriculture laws, municipal, and civil laws, customs, and sanctions. Pursuant to the 74th Constitutional Amendment Act, 1992, the municipalities have also been given the power to control water supply within the municipal limits. Again on the municipal use side, the Physical Health and Engineering Department (PHED) is responsible for ascertaining safety standards. For industries falling within the municipal limits, municipal corporation becomes the bulk supplier to the Industrial Development Corporation (IDCs), and outside the municipal limit, in industrial zones, the water resources departments supply water to IDCs. Thus, water governance is distributed and is under the control of various local level agencies, except pollution, which is regulated under a national law, the Water (Prevention and Control of Pollution) Act, 1974. Looked at scientifically, water is a complex hydrological entity and its availability and quality varies with regions, topography and climate. Given, its complex nature, water planning should be done by the national government, however, its administration, control and management should be left to the states, as is the case today. Even if seen from the climate change perspective, the action, if any, will have to take place at the local level, which can best be managed by the local governments. Further, it gives the state and local governments the power to make suitable adjustments in the regulations depending on the local needs.

Shirish Garud, Senior Fellow, Renewable Energy Technology Applications, The Energy and Resources Institute (TERI)

Water and energy are two vital inputs for economic development of a modern society. Impacts of anthropogenic activities on climate, commonly called as 'climate change', are now well accepted, and corrective actions to reduce and reverse these impacts are classified in two categories namely; mitigation and adaptation. Excessive use of fossil fuels as energy resources is identified as one of the major contributors to this. One of the major impacts of climate change is on availability of potable water in major parts of the globe. This makes studies of water, energy and environment nexus essential to develop effective mitigation and adaptation strategies.

Water, energy and environment nexus has multiple dimensions. While the energy and environment nexus has been studied in detail as a mitigation problem, the water-energy nexus is a resource and development issue, and water-environment nexus has a water supply dimension.

Water is usually considered as input for power generation in thermal and hydroelectric plants. Power sector, especially fossil-fuel based thermal power is one of the major contributors of GHG (Green House Gases) causing climate change. However, the same power is used for water extraction, distribution, and waste water treatment. Two common mechanisms for addressing water scarcity are ground water pumping and desalination. Both these mechanisms are energy intensive. Studies show that the requirement of power for water is going to go up in the coming future, as ground water levels are depleting due to excessive extraction, and surface water and water bodies are vanishing rapidly due to urbanization, and rapid changes in land cover, and land use patterns. This in turn will put strain on the climate. Hence water management policies must consider energy as an important factor. Both water and energy rank high on development goals but are rarely considered together. Agriculture sector in India accounts for around 86 percent of water used in India, and hence focus on water-energy nexus is of paramount importance.

Radheshyam Mopalwar, Director, Maharashtra Jeevan Paradhikaran

It is necessary that water is supplied and provided to the end consumer on volumetric basis, i.e. metered supply. Metered supply will ensure that NRW certainly gets reduced. This will help change emphasis from source development to efficient distribution, leading to better service. The experience of MJP is very positive in this regard, as all the urban water works that MJP manages have metered supply. Some of the Regional Rural Water Supply Schemes also have metered-watered supply. Metering, coupled with telescopic tariff, addresses the issue of equity effectively.

One of the most important issues is about the standards to be followed in metering. Even today, water meters across the globe do not have the facility of automatic meter reading (AMR). There is consistent pressure from vendors to adopt large scale AMR facility meters. The issue is of cost and is quite important at that. Today the per capita investment required in water supply schemes is about INR 6,000. Cost of an AMR meter is INR 12,000, excluding software and other system costs which are entirely hidden. The SITC (supply, installation, testing, commissioning) costs, plus 5 percent annual maintenance cost is different. The hidden costs include costs of the RF readers (which include a handheld reader costing INR 1,50,000, and a pole based reader costing INR 4,50,000), along with attendant costs, are not mentioned.

Now let us look at the cost of meter via-a-vis tariff. We collect INR 1,200 to INR 1,800 as annual water tariff from a 1/2 inch domestic connection. The cost of meter is to be borne by the consumer. It is impossible that a consumer will pay INR 12,000 towards measuring his supply of water, costing INR 1,200 annually. Compare this with tariff charged worldwide. It is equivalent to $400 annually. The international AMR meter price is $120. Despite such comfortable meter costs/tariff ratios, the world still has not shifted to AMR meters, whereas the meter costs/tariff ratios in India are heavily loaded against the consumer. Our insistence on AMR will kill the reform initiative. We need to press ahead with reforms and multi-jet EEC standard meters are good enough (cost INR 1200-INR 1500), or at the most ultrasonic meters with digital counters costing INR 3,000, where life of the meter needs to be guaranteed at 5-7 years (the operational life in case of conventional consumer meters has not exceeded 3 years in MJP).

Foreword

Much is expected of India in the years ahead. As one of the largest economies in the world, the general welfare of the global economy will be increasingly dictated and determined by events in India. Indeed, with China now facing a structural slowdown, with Europe mired in a deflationary trap, and with the US economy stuck on a subpar growth track, sustainable economic growth from India has taken on added importance. For the betterment of over a billion Indians, and the world at large, India needs to succeed.

But will it?

As this book goes to press India's policy makers are struggling to find the right monetary and fiscal mix for sustainable, non-inflationary growth. Like many other developing nations, India bounded out of the global financial crisis of 2008-09, emerging as one of the fastest growing economies in the world. The economy expanded by a robust 8.5 percent in 2009 versus a -0.4 percent decline in global output for the year. In 2010, India even outpaced China, with the economy expanding 10.5 percent versus China's 10.4 percent expansion. The difference in growth was slight but the symbolism of India hitting China-like growth rates was not. Early in the post-crisis environment, the general consensus was that India was finally poised to take off and assume the mantle from China in leading the developing nations forward.

Reality has turned out to be different, however. The global economic slump, juxtaposed against a number of internal structural barriers to growth has conspired to sap economic activity in India.

The economy expanded by less than 4 percent in 2012 and is expected to be only marginally higher in 2013.

The potential of India remains unrealized. Despite the nation's staggering political and economic success since gaining independence in 1947, much work needs to be done. Sustainable growth remains far from assured in India—and will go wanting if India fails to manage one of its most precious resources—water.

Rarely has a global commodity of such critical importance to economic development and prosperity received so little attention from policy makers. As we discuss in the following pages, virtually every economy in the world can survive without crude oil—considered by many as the globe's most important resource. But no nation can exist without water. Pick any sector of the economy—agriculture, energy, healthcare, transportation, technology—none of these key activities can thrive without a secure and sustainable level of water. Neither can humans, with the lack of clean water among the largest and most lethal health dangers in many parts of the world, India included.

Water is both the great enabler and disabler. It is critical to the success of any economy and society; it is among the most basic ingredients of growth. Hence it is also a great disabler. Without an adequate and secure supply of water, no society can thrive. While economists like to expound on the three essential ingredients of economic growth—labour, capital, and natural resources—water deserves to be mentioned in its own right. There is none more critical than water resource.

The challenge before India is to manage its water resources in a more effective and innovative manner. India's water crisis will only intensify if the government does not take concrete action to better manage existing and future water reserves. The nation's water challenges not only reflect a mismatch between supply and demand; it also mirrors the effects of weak policies and poor management on the part of the Indian government at all levels.

India's water challenges require leadership at the federal and local level, as well as leadership from the private sector and the special interests of agriculture, energy, transportation and a host of other sectors. In addition, meeting future water needs will require significant amounts of capital from both the public and private sectors; capital will also be needed from some key multilateral institutions like the Asian Development Bank and World Bank.

Yet another requirement lies with India learning from the rest of the world. The water challenge is global and hence various global responses have emerged over the past few years, with some best practices better than others when meeting the challenges of water scarcity. Effectively partnering, by both the centre and states on water management with select countries like China, Israel, Netherlands, Izech etc, poses a key challenge before India.

In the end, meeting the water challenge is an acute and strategic priority before India. As the following pages document, there is no greater policy imperative. Coping with rapid urbanization, rising industrialization, intensive agricultural use, soaring pollution levels, extreme weather patterns and the rise of a middle class—all of these challenges require water.

To sum it up, without an adequate and sustainable supply of water, India's economy is bound to underperform, if not stall in the years ahead. It is that simple.

The following pages shed some light on this all important and pressing topic. The nation still has time to avert a debilitating water crisis. But time is running short.

The demand for water investment in India is nothing short of staggering, which means the scope for investments in water is diverse and wide ranging. Just as India's water challenges are disparate and diverse, so are the investment opportunities. The nation's investment needs cut across various sectors of the economy—agriculture, industry, transportation, tourism, health-care and energy, to name just a few. They touch every sector of India's emerging 'green economy'. Meanwhile, the water challenge before India must be met by both the private and public sector—such is the importance of water to India's future. A public-private consortium is needed. National and regional coordination is also an imperative, as is closer cooperation on a regional and global scale. Key municipalities like Mumbai, Delhi and Kolkata—among the largest cities in the country and the world—should also be actively involved in mapping out India's water future.

The investment opportunities involving water run the gamut, ranging from waste water treatment, to desalination processing, to hydropower creation, to the deployment of watersheds and flood prevention policies. According to research from Merrill Lynch, the water industry could be worth $1 trillion by 2020. A large share of this growth will occur in water-scarce Asia, India included.

Joseph P. Quinlan,
Sumantra Sen,
Kiran Nanda

Rising Global Premium on Blue Gold

'Water has never been more under threat in modern history: pressure from a rising global population, industrialization, pollution and climate change itself are all putting fresh water supplies under strain.'

—*The Financial Times*, December 16, 2008

Viewed from up high—outer space—the world appears to be awash in water, and it is. Over 70 percent of the earth's surface is covered in water, with most satellite images showing a world wrapped more in dark deep blue swaths of moisture than brown streaks of land. That is the good news.

The bad news—roughly 97 percent of mother earth's water supply is salt water and therefore not fit for daily use. Of the remaining 3 percent, 2 percent, while considered fresh water, is locked in or frozen in snow and ice caps. That leaves around 1 percent for human use—or for a global population in excess of 7 billion, expected to rapidly increase to at least 8 billion in the next decade and a half. Without any doubt, two atoms of hydrogen joined to one of oxygen—water—is the world's most precious commodity.

Just how important is water? A recent intelligence report from the US government sums it succinctly:

'During the next 10 years, many countries important to the United States will experience water problems—shortages, poor water quality, or floods—that will risk instability and state failure, increase regional tensions, and distract them from working with the United States on important U.S. policy objectives.' (Excerpt from *Global Water Security*, The National Intelligence Council)[1].

In other words, the lack of water is a key risk to global stability. In the past, it was conflicts over oil that drove nations to wage war. In the future, the catalyst could be water, or the lack thereof.

Water, in other words, is the new oil—or an indispensible commodity that will either cause wars between nations in the future or bring the global economy to its knees, or both. An economy without access to water is like a human body without water—doomed, in dire straits, and at risk of a total collapse. The world in general and India in particular are rapidly approaching a tipping point in terms of water use and availability. Of the 'many countries' referred to above, India is one, highlighted as an at-risk country when it comes to having a predictable water supply.

Why Now?

How has something as mundane and plentiful as water become so important? Why are there siren-calls from governments and numerous multilateral agencies that the world is rapidly reaching a tipping point over water now? Is the world really running out of water?

Unlike the OPEC-led global oil shock of 1973—a solo event that dramatically altered the economics of oil for good—the emerging global water crisis is being driven by a convergence of forces. Greater urbanization, global climate change, soaring food demand and production in the emerging markets, a global population of over 7 billion people, underinvestment in the global water infrastructure, and the mispricing of water—all these variables have converged over the past few years to place unprecedented stress on the world's global water supply.

For the first time in modern history, just as many people live in cities as in rural areas, with rapid urbanization most evident in Asia, India included. Asia's cities are expanding by close to 100,000 people per day, one of the fastest levels of urbanization the world has ever experienced. It took 130 years for London's population to grow from 1 million to 8 million residents. However, Dhaka, the capital of Bangladesh, is expected to grow from 10 million people to

22 million people in just the next ten years. Meanwhile, the Asian Development Bank (ADB) projects that a staggering 1.1 billion people will move from villages to cities in Asia during the next 20 years—in their words, the migration is of a 'magnitude never before attempted by humanity[2].'

And it is not just more people living in the cities—it's about more people in general. As Al Gore notes in his book, *The Future*, the human population has quadrupled in the past century alone. By way of perspective, it took a staggering 200,000 years for human kind to reach one billion[3]. Now it takes just thirteen years to jack up the population by another billion. And while in 1990, only 13 percent of the world's population resided in the cities, by 2011 the percentage had jumped to 50 percent. In 2050, the figure will be closer to 70 percent. The key point—urbanization continues, and as this process gathers momentum, there will be even more demand on the world's precious water resources.

All of the above suggests that the world is becoming crowded and more urban-centric, a familiar story, too many in the developed nations. After all, roughly 82 percent of the US population lives in cities. Urbanization, in other words, is nothing new or unusual to most Americans, nor much of an issue in Europe (the rate of urbanization in Germany and Spain is 74 percent and 77 percent, respectively) or even central and South America, where nearly 80 percent of the population lives in an urban center. However, urbanization is still very much an ongoing process in Asia, notably India, suggesting more upside pressure on water resources here and throughout Asia.

What Happens When One Moves From the Farm to the City?

Globally speaking, urbanization is a big deal for the planet, representing one of the most significant global macro trends of our times. Life changes when you move from a mud or thatched hut in the rural areas to a shanty abode or a simple dwelling in the urban areas.

Rural living, for instance, is difficult, with many dwellers living without potable water and electricity, and dependent on kerosene, firewood or even cow dung for light and heat. They themselves or animals (water buffaloes, for instance), provide the energy needed to plough a small plot of land that provides a menial amount of food for a typical family. Schools are often windowless, open-aired facilities with minimal electricity. The children rarely commute to schools by buses or other transport. They usually walk, ride a bike, or hitch a ride on the back of the family's moped. Life in the rural areas, in other words, is rather simple, with the energy, food, and water footprint of a rural dweller often insignificant or not as intense as that of urban dwellers—outside the damage inflicted on the environment is right there at home.

However, things change when one moves from the farm to the city. For example, take a girl that once lived on a farm but moves to a major industrial centre to work in a manufacturing plant. Her migration sets in train a whole host of events that ultimately increases demand for energy, food and water. She will most likely live in company-sponsored housing that is powered by electricity and perhaps even airconditioned; if not, hundreds of fans are plugged into wall sockets, sucking on local energy supplies. She will use more water and eat more food, and as her income rises, she will of course send more money home and also have enough discretionary income to buy goods for herself—jeans, consumer electronics, global-brand beverages, make-up, hair conditioner and the like, all of which are water-intensive, i.e., it takes a great deal of water to produce these products.

These are the first steps and signs of becoming a global consumer. Take the one example from above and now multiply her by 200-250 million others just in China over the past decade, and you start to see how the transition from the rural areas to the cities can affect the world's water infrastructure.

Ultimately the girl earns money enough for a motor bike, increases her intake of protein, fruits and vegetables, and becomes hooked on western fashion, meaning more money spent on jeans, beverages, telecom and consumer electronics, fast food, and the like. All of these demands and products require additional clean water supplies. The girl is now placing even more demand on the global commodity structure, which is in stark contrast to her parents, grandparents, and great grandparents, whose rural-bound, meager existence for generations keep them on the fringe of the global economy and at best, a nominal source of demand on the global water system.

Today, roughly one in two people in the world live in a city, with that figure expected to jump to two out of three in the next few decades. Meanwhile, more city dwellers mean more pressure—intense pressure—on the physical water infrastructure of many nations, many of which were already under severe water constraints before the current mass migration. Water usage soars as the number of urban dwellers increases, with urbanization not only placing additional demands on the existing water supply but also raising the stakes to find and secure new supplies of water. The growing thirst of booming cities in Asia has been so dramatic that the ADB has called for a major rethink on how countries in the region develop and build cities in the face of tightening water constraints.

And finally, let's not forget that rising per capita incomes in the cities also translates into more human waste. According to the World Bank, the per capita production of garbage from the world's urban residents is now 2.6 pounds per person per day[4]. That is a staggering statistic but it gets worse—based on World Bank's estimates, the total volume of urban garbage is expected to increase 70 percent over the next dozen years[5]. And where does a great deal of this garbage end up? It ends up in many streams, rivers, lakes and oceans in both the developed and developing nations. Presently, more than half the world's rivers are seriously depleted or polluted.

The Growing Thirst of the Rest

One reason behind the soaring premium on water lies with the fact that for the first time in modern economic history, the rest of the developing nations, are at the forefront of driving global growth, placing unprecedented demand for natural resources in the process.

The global economy no longer beats to the tune of the West—or the small populated, but wealthy cohort of the US, Canada, Japan, Australia, New Zealand, and developed Europe. For years, developed nations set global commodity prices, while the 'rest' were price takers. That was the past. The future will be different.

Indeed, it is the developing nations that increasingly exert the most pressure on the world's natural resources. In the search for commodities, water included, the West has failed to recognize the critical fact that it is now in direct competition with the 'rest' for natural resources, notably water. In other words, the long-standing monopoly the West has enjoyed in devouring the world's natural resources is over. In years and decades past, as long as consumers in the developing nations remained poor and lacked the income to purchase a computer or car, or afford a good meal, the West did not have to compete with the developing nations for oil, copper, water and other commodities. For much of the post-cold war era, the equation was rather simple: the developing nations produced commodities, the West consumed them. Those days are past, however. The budding new middle class of the emerging markets has changed the equation.

How big is this middle class cohort? According to the World Bank, the middle class of the developing nations is relatively small, yet, nevertheless poised to expand rapidly over the next few decades[6]. This group already accounts for some 400 million people, according to the Bank, a figure roughly one-third larger than the entire US population. More importantly, the middle class of the developing nations is expected to triple in size over the next two decades, increasing to 1.2 billion shoppers by 2030. By then, the developing nations will account for 93 percent of the global middle class, up from 56 percent in 2000[7]. This estimate suggests ever rising levels of water consumption from the developing nations.

Most average Americans are oblivious to the rising middle classes of the developing nations and what this new consuming cohort means for the world's already stretched natural resource base. They have yet to recognize that as the new global consuming class adopts and acquires western lifestyles—moves from the village to the city, works in air-conditioned offices, drives to work, consumes more protein—more the demand, higher the prices for energy, water, agricultural goods and other natural resources.

Given the demand and supply conditions, the price of water will increasingly reflect the rising per capita incomes and the attendant jump in consumption among consumers in the developing nations. Thanks to the rising consumption among emerging market consumers, later, between now and 2030, worldwide demand for food is expected to rise by 50 percent and demand for meat will jump 85 percent, according to the World Bank[8]. In that, agriculture is a huge source of water demand—agriculture uses approximately 70 percent of the global fresh water supply—the more these consumers consume meats, fruits and vegetables, the greater the demand on the world's water infrastructure. Think rising global water scarcity or more conflicts between states as the world scrambles to secure a more reliable water supply. Related to these trends—ever rising levels of saline and pollution of key water bodies and aquifers, and the continued degradation of the world's water supply.

As part of the new world we live in, the competition for resources, with water forming the front and the centre of this conflict, that could very well lead to more tension between the well-endowed 'rest' versus the depleted, resource-deficient West.

To revert to the US intelligence report on global water security, the report highlight the following[9]:

- *'We assess that during the next 10 years, water problems will contribute to instability in states important to U.S. national security interests. As a result of demographic and economic development pressures, North Africa, the Middle East and South Asia will face major challenges coping with water problems.'*
- *'We judge that as water shortages become more acute beyond the next 10 years, water in shared basins will increasingly be used as leverage; the use of water as a weapon will become more common during the next 10 years. Water terrorism could become a major issue for some nations.'*
- *'We judge that during the next 10 years the depletion of groundwater supplies in some agricultural areas will pose a risk to both national and global food supplies. There is a strong correlation between water available for agriculture and national GDP.'*
- *'We judge that, from now through 2040, improved water management (e.g., pricing, allocations and 'virtual water' trade) and investments in water-related sectors (e.g., agriculture, power, and water treatment) will afford the best solutions for water problems.'*

All of the above makes clear that the world—distracted by one financial crisis after another, over the past few years—is heading for a crisis which is of far greater magnitude.

Today, nearly one in every four human being lives in areas of physical water scarcity on account of inclement weather, water-waste and sub-par agricultural practices. The water infrastructure in the US and overseas is crumbling while demand is rising. According to a few mainstream forecasts, one in three people in the world will face some type of water shortage by 2025. Presently, one out of eight people lack access to clean water, while in fifteen years 1.8 billion people will live in regions of severe water scarcity. Around 46 percent of the people in the world do not have water piped into their homes. Some 2.5 billion people in the world do not have access to a toilet, equivalent to 40 percent of the world population. And the disparity of water usage is massive: for instance, while Americans use about 100 gallons of water at home each day, millions of the world's poorest people subsist on fewer than five gallons. Accordingly, women in the developing countries walk an average of 3.7 miles to get water. Gabra women in northern Kenya spend up to five hours daily hauling water.

Drought conditions have become more severe in many parts of the world, while the pace of global urbanization has accelerated, straining the water infrastructures of many nations to their breaking points. Population growth and urbanization are expected to drive demand for water up 40 percent within twenty years according to the World Bank[10]. Then there is the unrelenting demand for 'virtual water'.

The Soaring Demand for 'Virtual Water'

Virtual water is the amount of water used to create a product; it's the water consumed indirectly when making a pair of jeans, a cotton T-shirt, or the amount of water needed to produce a pound of meat, processed cheese, or fruits and vegetables like plums, cherries and bananas. These are every day products that require, unbeknownst to many, a staggering amount of water.

It takes, for example, 2,900 gallons of water to produce one pair of blue jeans; a simple cotton bedsheet requires 2,800 gallons of water. One hamburger requires 634 gallons of water, while just one cup of tea requires 9 gallons.

The water intensity of producing one pound of beef is not insignificant, requiring 1,857 gallons of water. A pound of chicken requires nearly 500 gallons of virtual water. Regarding other products—sausage–1,382 gallons; a pound of processed cheese–589 gallons, and a pound of bananas–103 gallons.

All of the above strongly underscores the fact that the direct and indirect demand on the globe's water infrastructure is staggering.

Nearly every economic activity requires some water intake, making the water the most precious commodity in the world. In the content of 'virtual water', the underlying global demand for water is truly staggering and set to only intensify as more and more middle class consumers in India and elsewhere place unrelenting demand on the world's global infrastructure.

In short, the direct and indirect demand on the globe's water infrastructure is simply stunning. Nearly every economic activity requires some water intake, making water the most precious commodity in the world.

Regional Stress Points Continue to Build

The demand for water keeps building throughout the world. In the Middle East, for instance, rapid population growth and water shortages are colliding in a region where the population is expected to double to over 600 million during the next forty years. Future water shortages may play a more prominent role in social unrest throughout the region. According to the World Bank, Sana'a, the capital of Yemen and a city of two million people could run dry within the next six years[11]. Aquifers that can be replenished are being pumped faster than their rate of recharge. Of the nation's 21 main aquifers, 19 are no longer being replenished.

In China, according to ADB, with only 7 percent of the global water resources and almost 20 percent of the world's population, the per capita water volume is only one fourth of the world average[12]. More than half of China's cities suffer from water shortages, affecting more than 160 million people. Unequal distribution of water further exacerbates this situation. For example, the northern region is home to 40 percent of the population, but only 14 percent of the water.

According to the World Bank, water shortages cost the country roughly 1.3 percent of its annual economic output, plus another 1 percent is lost to water pollution. Against this backdrop, China has taken the unpopular step of raising water prices, with the city of Shanghai boosting residential water prices 25 percent in June, with plans for additional water hikes in the future. At least half a dozen other major cities have raised water prices the past few months, moves that have left Chinese consumers angry and upset with the decline in their purchasing power.

That said, water prices in China are relatively low on a global basis, although the recent hike in prices—with more likely in the future—could emerge as a head-wind to household disposable income and consumer spending.

There is a very real and grave risk that the great Chinese story—the next alleged world superpower—could be subsumed and derailed by the world's most precious commodity, water. Should this happen, the ramifications will be felt not just in China but around the world.

In Africa, meanwhile, water scarcity and sanitation are both major challenges; simply put, if the continent fails to secure a reliable supply of water, the region is destined for failure. As of now 14 African countries are subject to water stress or water scarcity and an additional 11 will join them by 2025 according to a study by the United Nations Environmental Programme[13].

In the Middle East, among the driest places in the world, the Jordan River has been largely reduced to a trickle, the surface of the Dead Sea has shrunk by one third, the Sea of Galilee is at its lowest point ever, and in Iraq, the Euphrates river is drying up, possessing a host of economic problems to the war-torn nation.

Meanwhile, as reported in the *New York Times*, 'In Northern Syria, more than 160 villages in the past two years have run dry and been deserted by residents. In Gaza, 150,000 Palestinians have no access to tap water. In Israel, the pumps at the Sea of Galilee (Lake Kinneret), its largest reservoir, were exposed above the water level, rendering pumping impossible. In Lebanon, 70 percent of wastewater is dumped into ground water-polluting cesspools and Jordan is struggling with just 10 percent of its average rainfall[14].'

Against this backdrop, water scarcity represents a massive economic and geopolitical challenge to the Middle East.

The lack of water is also a significant challenge in the US. Parts of Texas are struggling with one of the worst droughts since World War I, a two-year dry spell costing the agricultural sector billions. The forced sales of cattle, dwindling cotton yields, a steep decline in shellfish production, and lost tourist revenue—just some of the economic aftershocks and consequences that have accompanied one of the worst droughts in the history of the lone star state.

In Colorado, meanwhile, harvesting rainwater is now legal given the premium on water, while in Arizona, water harvesting has become a huge business. California, has recently seen the approval of the first large scale desalination plant in San Diego, with more approvals expected as one of the largest states in the union grapples with the scarcity of water.

The Mounting Water Crisis in India

In India, another developing nation expected to emerge as a global economic growth leader in the next decade, there are worrying problems too. Thought by many to be one of the most dynamic economies among the developing nations, India's economic potential could very well get derailed or aborted by the nation's looming water crisis.

The economy, one of the largest in the world, is overly dependent on water, or to be more precise, monsoons. A disappointing monsoon season can wreck havoc on the pace of growth in one of Asia's largest economies. Given a severe shortage of proper irrigation facilities, the lack of rain is notably difficult for India's agricultural sector, accounting for nearly one-fifth of its gross domestic product. More importantly, roughly two-thirds of India's population relies on farming and related industries. Hence, poor monsoon rains have a huge negative multiplier effect on the rest of the economy.

In particular, the lack of rain can easily and often trigger a hike in food prices, a dynamic that erodes the purchasing power of both urban and rural households. In Mumbai, meanwhile, water rationing is now in place, a step that could slow economic growth in one of India's most important cities. While the ADB recently forecast that India will confront a serious water crisis by 2020, many believe the crisis will arrive a great deal earlier[15].

Indeed, India is a country under intense water stress. Consider the following:

- India currently has the world's second largest population, and is expected to overtake China's by 2050, when it is likely to reach a staggering 1.6 billion, putting unheard of stress on the available water resources. Presently, the nation has 17percent of the world's population, but only 4 percent of its usable fresh water
- The per capita availability of water in India has dropped from 5,300 cubic meters per annum in 1951 to 1,544 cubic meters in 2011. Thus, the country is already in a 'water stress' situation
- Water is under growing pressure both on the supply side (insufficient freshwater, uneven distribution, poor quality, non-revenue water, climate change), and the demand side (agriculture, industry, residential). Moreover, even the reduced availability of fresh water is highly uneven over time and space
- Besides, the declining per capita availability of water, there are also issues of deteriorating water quality
- According to the World Watch Institute, India suffers from misgovernance on the water pricing front—giving either water free or at heavily subsidized rates. This naturally has led to over-exploitation of ground water and widespread environmental damage

- Water sector is crucial for the Indian economy on account of its growing population and production of goods and services. Water shortages imply a future of food shortages, as well as other essential goods and services shortages
- For years, experts have been issuing warnings about the likely onset of water wars. However, the policymakers have failed to realize the gravity of the situation
- Indian economy has started witnessing the first glimpses of conflicts over water, like many countries in the Middle East, Africa and Central and South Asia. Competition for water is leading to social unrest, conflict, and migration
- India's 12th Five Year Plan's (2012-17) thrust is on providing a facilitative environment for the amelioration of water related problems. Time is of essence, however. The country can no longer postpone concert actions to address its water crisis

All of the above outlines the stark challenge in front of India. The problem is complex—how to provide clean water to a population of over one billion people who are eating better, driving more, and living in urban areas. The solutions are just as complex and expensive—it will take billions of dollars in investment to head off this crisis. The outcome, however, assuming nothing is done in the immediate future, is rather simple: India will not only fail to grow and prosper but will see its economic development and prosperity retard and diminish. Economic uncertainty—if not economic and social chaos—will ensue if India does not meet its water challenge.

Solutions

Against a backdrop of global water scarcity, capital expenditures on building and improving water filtration, waste water treatment, desalination, rural water services, water utilities, water utility performance, irrigation systems, and a host of related projects are poised to increase sharply during the next decade. Trillions of dollars are likely to be spent by governments and corporations over the next few decades, a necessary and imperative capital expenditure lest many economies falter and stagnate. The choice is that simple and stark.

The capital has to be spent or countries like India and China will simply not grow. Indeed, their economic development will be reversed; their societies and populations sent reeling backwards instead of forward. In fact, it is these threats and potential solutions for India that form the topic of the following chapters.

In brief, we expect infrastructure expenditures to materialize via public/private partnerships. The Global Water Partnership estimated in 2000 that global water infrastructure requirements would amount to $180 billion per year through 2005—actual annual private investment over the period totalled $3.3 billion, or just 1.8percent of the forecast. Lack of investment has sent fixed water infrastructure into disrepair in many of the world's fastest growing cities. Moreover, according to the World Bank, private financing has accounted for less than 10 percent of total water supply and sanitation investment in the developing world to date[16]. With existing water infrastructure strained to the point of capacity, and many emerging market governments flush with capital reserves, thanks to booming trade surpluses, we suspect that future infrastructure build outs will rely on the expediency of private sector expertise and the depth of public sector capital.

Water: The Great Enabler and Disabler

Virtually every form of life pivots on water. Every sector of the economy, and every economic activity, is dependent on water, whether the industry is agriculture, mining, energy, retail, construction and other related activities. Farmers need water to farm. Mining firms need water to mine. Energy firms need water to create energy. And households need water just to function properly. Water affects the air we breathe, the food we eat, our health, and the energy we consume. Water, in short, is the great enabler, the most important ingredient in the world when it comes to the growth, peace, and prosperity of nations.

By the same token, water—or the lack thereof—can also be a significant disabler, or a growth-stopping impediment to development and a key source of political instability. As the US intelligence community notes, while water issues by themselves are unlikely to result in state failures, water scarcity can nevertheless contribute or exacerbate poverty, social tensions and inequalities, environmental degradation, and failed institutions.

Water related issues can also lead to state-to-state conflicts. Over fifteen developing nations generate 80 percent or more of their electrical power from hydropower, the lack of water in these states would lead to a sharp fall in electricity usage, a steep decline in economic growth and increased border tensions, in many cases, with neighbouring nations. To the latter point, India and other nations are worried about China's 'hydro-supremacy' on the continent, particularly with regards to China's plans to build huge dams on rivers that flow across international borders.

Water scarcity is also emerging as a key source of internal tension for many nations. The debate about whether to allow gas and oil hydraulic fracking in the United States has triggered a rift between environmentalists on the one hand, and against energy extraction companies on the other. So-called 'fracking' is a process in which water is mixed with sand and chemicals and then injected at high pressures in horizontal and vertical beneath the ground. The technique is very water intensive. Hence, mining and energy companies all over the world are up against daily resistance to rising water usage of their operations.

In addition to the above, water is a matter of life or death. As the World Health Organization notes, better water and better sanitation could prevent 6 percent of all deaths in the world—children and adults. Among the diseases that thrives on the lack of water sanitation is diarrhea, which takes the lives of over 4,000 children under the age of five each day, according to the United Nations[17].

Water is the great enabler. But for many nations, India included, it is rapidly becoming a disabler or a significant threat to growth and prosperity.

Running Out of Time

The following quotation aptly applies to the current state of thinking on water.

'We have not succeeded in answering all our problems. The answers we have found only serve to raise a whole set of new questions. In some ways we feel we are as confused as ever, but we believe we are confused on a higher level and about more important things.'

—BERNT OKSENDAL

The population of India is around 1.21 billion (2011 Census), which accounts for about a sixth of the world's population. This translates to 17.5 percent of world population while the growth rate is pegged at 1.41 percent[1]. Statistics depict that more than 50 percent of India's population is below the age of 25 and more than 65 percent below the age of 35 years. Predictions for the next decade portray that the average age of an Indian will be 29 years old[2]. The increasing youth population can prove to be an asset for the country in the near future, as the youth bulge seems to be one of the sources of future economic growth in India[3]. India's demographic dividend by itself is likely to contribute two percentage points to its annual Gross Domestic Product (GDP) growth for the next two decades, provided the country adopts the right policies[4]. Right policies mean that policy designers need to have the perspective of dovetailing investment and development plans in all three sectors namely services, manufacturing and agriculture. All policies, aiming to boost GDP in the coming years, will have to be integrated with water policies as water is a key resource for the very basic as well as the most sophisticated needs of modern society. Growing population will demand never-ending supplies of food, clothing and space, as basic necessities, availability of which cannot materialize without water. India's rising population and its high trajectory of economic growth, notwithstanding recent slowdown, has placed tremendous pressure on India's existing water resources.

Demand for water is steadily increasing across all markets and sectors, and will continue to do so.

Balancing water demand with the available supply will be the key determinant for future economic growth and sustainable development. The supply of water is largely dependent on seasonal showers that are characteristic of the Indian subcontinent. In India, rainfall pattern has become the biggest challenge for water harvesting and water management. Around 80 percent of annual rainfall is received in the duration of only 5-6 months accompanied by varying patterns. Moreover, due to global warming, it is observed that Indian monsoon has become increasingly unpredictable and unreliable. In addition to this, the increasing frequency of extreme events such as floods and droughts over the last decade or so have been a major contribution for breakdown and disorientation of water supply. While availability of water is limited, the demand for water will continue to increase at a rapid rate due to growing population, haphazard urbanization, rapid industrialization and economic development.

Therefore, availability of water for utilization needs to be supplemented with efforts to satisfy increasing demands of water. Direct use of rainfall and avoidance of inadvertent evapo-transpiration[5] are the additional strategies that should be adopted for augmenting utilizable water resources.

India receives an average annual precipitation of about 4,000 billion cubic meter (BCM), which is its primary source of water. After consideration of the natural evapo-transpiration, only about 1869 BCM is the average annual natural flow recorded through rivers and aquifers, of which only about 1123 BCM is useable through the existing strategies, if large inter-basin transfers are not accounted for. The Central Water Commission (CWC) has assessed the total utilizable water resources of the country to be 1108.849 BCM, which includes 690.309 BCM of surface water and 418.540 BCM of ground water. Another area of concern is the consistent depletion of water table. The annual extraction of ground water in India is over 150 BCM, which is one of the highest in the world. Thus, there is need to arrest the declining ground water levels in the over-exploited areas. This can be best done by introducing improved technologies of water usage. Also efficient use of water can be motivated by incentives and by encouraging community based management of aquifers. In order that recharge is more than extraction, artificial recharging of projects should be undertaken. This would enable the aquifers to provide base flows to the surface system, and thereby maintain the ecological balance.

With a crumbling water infrastructure that prevents the government from being able to supply clean drinking water, millions of Indians are left with zero or minimal access to drinking water. In India, only 43.5 percent of households use tap water, while 87 percent still depend on the tubewell, handpump and covered-well as the main source of drinking water, according to the Houselisting and Housing Census 2011[6].

According to data released by union home secretary R.K. Singh, only 47 percent of households have source of water within the premises while 36 percent of households have to fetch water from a source located within 500 m in rural areas and 100 m in urban areas.

As many as 17 percent still fetch drinking water from a source located more than 500 m away in rural areas, or 100 m in urban areas.

Thirsty, Haphazard Urbanization

Urban water demand is rapidly growing in India due to high growth in its population, which is led mostly by migration and rapid industrialization. Meeting this demand is one of the biggest challenges that urban planners face. Incidentally, large urban areas experiencing rapid growth in population are in arid and semi-arid regions, which are topographically water-scarce. As a result, water supplies from local water resources, including aquifers, are grossly insufficient for the high and concentrated demands in most urban areas. Under such circumstances, these large cities have to rely on distant reservoirs that have considerable capacity and are self-sufficient. The analysis of 302 urban centres shows that cities with large population have a much higher level of dependence on surface water sources. Also, greater the share of surface water in the city water supplies, higher is the level of per capita water supply[7]. Urban population has doubled over the past thirty years and accounts for nearly 30 percent of the total population as of 2009. It is expected to reach 41 percent by 2025, to over 575 million people from the present level of 286 million[8]. The growing metros/cities like Delhi, Mumbai, Bangalore and other Tier I cities are attracting a significant number of commercial and institutional establishments. Every year, there is a growing number of large and small offices, IT parks, malls, hospitals, schools, colleges, hostels and hotels that operate in these cities. Subsequently, there also is an increase in the usage of amenities such as flush toilets, showers, washing machines, cooling plants and other water intensive facilities, which essentially dwell in a typical urban habitat. Indeed, city dwellers lead a more water intensive existence.

Big and developed cities have become role models for the medium or small Tier II –Tier III cities. With growing prosperity, these cities unknowingly follow the same trend of haphazard urbanization. As more and more people and businesses migrate in cities, water demand is bound to accelerate. The situation goes on worsening, resulting in increasing intra-city inequity and unbalanced growth patterns. To accommodate this migration and consequent agglomeration patterns, the cities tend to ignore their green belts, wetlands, rural hinterlands and other natural micro-environments, which are generally perceived as city's breathing spaces (for example, mangroves, and forest land in the western and central region of Mumbai). This, to a large extent coincides with growth of the construction sector in India, which is witnessing a boom similar to the IT industry in the nineties[9]. In 2005, the official water demand for India's largest cities of Delhi and Mumbai was pegged at an unimaginable 3,973 million litres per day (MLD) and 3,900 MLD and the per capita demand was estimated at 268 litres per capita per day (LPCD) and 307 (LPCD) respectively. However the water supply often falls short of the ever increasing demand. As a result, these cities are constantly facing demand-supply deficit which most of the times leads to conflicting situations. It has been observed that the shortfall in such cases was about 600 MLD and 900 MLD in Delhi and Mumbai respectively. Most Indian cities, including Chennai and Mumbai, depend on annual rainfall for their yearly water supply. Chennai has been suffering from water shortage for decades. The requirement for the city and the adjacent areas is around 1,470 MLD, which includes commercial and industrial demand. But the city gets daily supply of only 600 MLD from sources such as lakes and reservoirs, which are dependent on the erratic monsoon patterns. Chennai now has a desalination plant, adding to its demand by another 100 MLD. The situation is assuming critical proportions, due to the fact that, almost 30-40 percent of water is lost during transmission, and supply in almost every city in India, which is disgraceful. As a result, cities are on a constant expedition for augmenting supplies by tapping distant water sources which are originally intended for rural areas or some other uses.

Besides increasing population, the rising water demand is also attributed to the changing lifestyle and consumption patterns across the country, especially in the urban centres. Although there are varying standards and estimates of average water use per capita per day made by different agencies, but there is hardly any argument about the fact that toilets and bathrooms are the biggest water guzzlers in a urban house, with flushes, taps and showers devouring more than 60-70 percent of total water use[10]. Innovative usage of water, for instance, recycling of water used in showers and washing machines for flushing the toilets can be another option.

Rural and Irrigation Outreach

Rural India has provided safe drinking water facilities to more than 90 percent of the rural households, according to the National Sample Survey Office (NSSO) 65th round survey 2008-09[11]. This was achieved as a major step forward through active government policies towards rural drinking water provision, which came in the wake of the disastrous Bihar famine of 1967. The subsequent years saw increasing policy efforts at the designing and the implementation level, to address the issue of water scarcity in 'problem villages' all over the country. The generic method for the most basic rural water supply systems are dug wells, tubewells, rainwater harvesting, and pond sand filters. With arrival of the new technology of drilling rigs and new hand pump models in the early 1970s, the objective of national and state governments was reduced to finding funds to implement the 'formula' of piped water supply schemes, wherever viable. The Accelerated Rural Water Supply Programme (ARWSP) in 1972 and the short-lived Minimum Needs Programme of 1974 (replaced by the ARWSP in 1977), were efforts to target increased government outlays for villages with continued water shortages and water quality problems[12]. This mode of water-service delivery characterized the initial expansion and mission phases and the ARWSP, which constitutes the bulk of funding for rural water supply, continues to be implemented through this mode in several parts of the country with satisfactory results. The traditional mode of water-service delivery was a supply-driven provision designed and implemented by the engineers (rural drinking water supply programmes run by the state and central government) of the Rural Water Supply (RWS) or Public Health Engineering Departments (PHED), depending on the institutional setup in each state.

By the end of the 1990s, three major surveys—the National Commission on Water (NCW, 1999), the India Water Partnership report (IWP, 2000) and a six-volume review of India's water sector by the government of India and the World Bank[13] (GOI-World Bank, 1999)—concluded that there was an emerging water crisis in India's water sector. Further, the World Bank study said, 'India faces an increasingly urgent situation; its finite and fragile water resources are stressed and depleting while different sectoral demands are growing rapidly[14]. Water is becoming an increasingly scarce resource in India, yet it continues to be used inefficiently on a daily basis in all sectors, while sectoral demands (such as from drinking water, industry, agriculture and others) are growing rapidly in line with urbanization, population increases, rising incomes, growing aspirations and industrial growth. At the same time, the poor and disadvantaged remain underserved by the heavily subsidized public services, and have to bear increased health risks plus additional costs (in terms of time and money) of obtaining potable water supplies. Women and children are disproportionately affected under these conditions due to their greater role in water collecting activities[15].'

It was through the World Bank report, that for the first time, the matter of sustainability of drinking water supply came under the public lense, as an issue that could soon turn into a major crisis. Among the 170 recommendations of the joint government of India and World Bank review (GOI-WB, 1999), there was an important recommendation which suggested the introduction of a demand-responsive, community-based approach to the provision of rural water supply. From the 1970s, non government organizations (NGOs) have been piloting such approaches in different parts of the country, including Gram Vikas in Orissa, in Karnataka and Utthan and SEWA in Gujarat. More recently, during the 1990s by partner agencies of WaterAid in Tamil Nadu, the DANIDA-supported Rural Water Supply Project in Tamil Nadu, the Socio-Economic Units Foundation (SEUF) in Kerala, and the World Bank-supported Swajal project in Uttar Pradesh, threw up at least five characteristics of successful rural interventions in drinking water supply that were at odds with the government's approach: (1) community cost-sharing, participation and ownership; (2)a focus on reviving and maintaining traditional water-harvesting structures; (3)demand-responsive approaches attuned to community needs and priorities; (4)social equity to ensure equitable coverage within the entire community, particularly the inclusion of poor and other disadvantaged groups; and (5)placing women at the centre of water management decisions. Yet, none of these were implemented as an integrated approach, and individual NGOs focussed on one or more aspects based on trial and error.

Irrigation, if explained in economic terms, is arguably the most essential input in the agriculture production function. It plays a crucial role by substantially complementing the production process. The other key inputs, namely, seed and fertilizer cease to realize their full benefit unless substantial irrigation facilities are available[16]. Irrigation sector uses water the most and presently accounts for 84 percent of the total water withdrawals[17]. Unlike in most developed countries, agriculture in India (and other similar countries) is characterized by the predominance of small farmers (farmers with small landholdings, typically less than two hectares), majority of who depend on rainfall for agriculture[18]. About 65 percent of the net cultivated area is rainfed, which makes agriculture and related livelihoods mostly vulnerable to shifting rainfall patterns[19]. The irrigation sector has been fundamental to India's economic development and poverty alleviation since 25 percent of India's (GDP) and 65 percent of employment is based on agriculture.

During the post-independence period, the country made huge capital investments in many major and medium irrigation projects, and with the advent of the drip irrigation system, there has been a modification with respect to the consumptive use of water in agriculture. For the maintenance of irrigation systems in an efficient manner, there are huge operations and maintenance (O&M) costs involved which fall within the government's purview. The annual budget releases for O & M of major projects are grossly misquoted and inadequate. In some instances, under the pretext of economic dealings, attempts have been made to prune down the lower level staff deployed in the O & M of major projects. This has resulted in forcing the administration to shell a greater amount within every few years to repair the system and put it back into order. Indirect losses, by way of less crop output, and others from improper maintenance of the system can also be sizeable[20]. Around 2006, CWC reported that the water fee realized by all major and medium irrigation projects was equivalent to 8.8 percent of the 'working expenses' during 1993–97 and the ratio had declined further to 6.2 percent during 1998–2002 (CWC, 2006) compared to 2.5 to 3 times of water expenses around 1900.

Industrial Water Consumption

India is one of the emerging urban economies in the world. The growth is characterized by a specific shift in contribution to GDP from agriculture to tertiary and manufacturing sectors, thus bringing urban areas into the centre stage of development process. It is estimated that roughly, 60 percent of India's GDP is generated in urban areas[21]. The industrial sector in India is the second highest consumer of water after agriculture though there is minimal quantitative evidence with respect to data that can help gauge the actual magnitude of industrial water use and impacts; especially its impacts on sustainability of agricultural growth[22].

Availability of water is soon turning into an area of deep concern for the industries and their expansion plans. While the provision and management of water has typically been a responsibility of the government, a paradigm shift around water has emerged, which focusses on the concept of corporate water risk[23]. In 2011, a leading national chamber's water mission undertook a survey with its member companies to gauge the importance Indian companies attach to water, its conservation and management; the study points out that with regard to the current availability of water, while 60 percent of the respondents agree that availability of water is impacting their business today, the figure rises to 87 percent ten years hence. Member industries have acknowledged the fact that over the past few years, access to water has become difficult and the problem is likely to increase in the coming years. This is a matter of grave concern for industries belonging to sectors like thermal power plants, chemicals, textiles, cement and manufacturing[24]. While the need for better water management is well known, there is a lack of reliable and quantifiable information on water consumption at the enterprise level and almost no consensus on the range of industrial water demand by the various sectors and their respective price elasticity.

The nature of risk in different socioeconomic and geographic settings has not been articulated beyond a first order analysis of the potential imbalance between estimates of average annual supply and demand[25]. A study conducted by FICCI and Columbia Water Center (CWC) emphasized that, at least one in five multinationals in the most water-intensive sectors is already experiencing slowing of growth plans resulting in balancesheet losses in their businesses due to drought and other shortages, flooding, and rising prices. Thus, even as businesses seek to secure long-term prosperity, to maintain competitive advantage and brand differentiation, and to secure stability and choice in supply chains—depending on the type of business there will be different levels and types of risks related to increasing scarcity of water[26]. Findings suggest that ground water is the major source of water for different industrial sectors across India. Some 55 percent of those surveyed used ground water with or without some other source of water. Surface water, with or without another source, accounted for 51 percent of water sourcing, while municipal water, with or without another source accounted for 44 percent of water sourced[27]. Observations imply that, environmental changes also pose a risk but are definitely not perceived as a 'severe' risk, even though a larger percentage feel the pinch of inadequate water availability and 87 percent cite this factor as the predominant risk in the next ten years. This exemplifies what may be best described as a lack of awareness on these matters amongst industry members. Allocation policy and pricing—which are tangible, affect the 'present', are seen as potential risks, but not that severe[28].

In the draft National Water Policy (2012), it is suggested—'Industries in water short regions may be allowed to either withdraw only the makeup water or should have an obligation to return treated effluent to a specified standard back to the hydrologic system. Tendency to unnecessarily use more water within the plant to avoid treatment or to pollute ground water needs to be prevented[29].'

Solution to the above dilemma—whether water is a public good or an economic product has been attempted in the National Water Policy (2012) as a part of the 12th Five Year Plan. It inter-alia indeed recommends that for the pre-emptive and high priority usage of water for sustaining life and ecosystem so as to ensure food security and supporting livelihood for the poor, the principle of differential pricing may have to be retained. Over and above these uses, water should increasingly be subjected to allocation and pricing on economic principles.

'The principle of differential pricing may be retained for the pre-emptive uses of water for drinking and sanitation; and high priority allocation for ensuring food security and supporting livelihood for the poor. Available water, after meeting the above needs, should increasingly be subjected to allocation and pricing on economic principles so that water is not wasted in unnecessary uses and could be utilized more gainfully.' **(National Water Policy 2012)**

Although the intent of pricing water right seems to be in place, what needs to be prioritized is how the proposed system translates the same into practice. As India is not the only nation caught in this debate and state of confusion, it can take cues from some nations who have or are now trying to look at the issue right in the eyes. Many of these policy and change makers are aware that increases in prices will be—and in some countries are already proving to be–hugely controversial, they feel that they can be managed without negatively impacting the poorest. They also concur that as long as most countries provide huge subsidies for water, it will not be possible to change the wasteful habits of consumers, farmers and industry, nor to raise the investment needed to repair old supply systems and build new ones.

Still, there is a lurking ambiguity whether water is to be treated as state property or private property or a common pool resource. The document, on the one hand, contains statements about water being a community resource and about the public trust doctrine being applied to its usage and on the other, there are references to water as an economic good and also to public-private partnership.

On an urgent basis, collaboration with select countries in various forms is required to make Indian water sector efficient and viable. Already some countries like Israel and Netherlands have entered into specific collaborations with some states and the centre. The existing framework enables and supports the realization of community good, as well as affords opportunities for realizing equitable distribution of earth's resources, including water.

To conclude, it is important that innovative thinking should reflect itself first at the policy-framing exercises as India's water woes are on its way to assuming mammoth proportions. With its rising young population and finite resources, India lives closer to the edge of survival than many other countries. It has less play and less time to get water sustainability right.

Appendix

Annual Requirement of Water in India (1990, 2000, 2010, 2025 and 2050)					
					(In BCM)
Different Uses of Water	**Year**				
	1990	2000	2010	2025	2050
Domestic	32	42	56	73	102
Irrigation	437	541	688	910	1072
Industry	-	8	12	23	63
Energy	-	2	5	15	130
Others	33	41	52	72	80
Total	502	634	813	1093	1447

Source: India Stats

Future Drinking Water Demand in India (1991, 2001, 2011, 2021, 2025 and 2050)					
Year	**Total Water Demand**		**(Billion Cubic Meter) BCM/Year**		**BCM/Year**
	Based on Past Census	**Based on UN Projection**	**Based on Past Census**	**Based on UN Projection**	**Total Water Requirement for Domestic use for Urban and Rural**
1991	31465		11.48		
2001	43065	49935	15.72	16.03	
2011	54810	63555	20	23.2	42
2021	66555	83375	24.29	30.43	NA
2025	71340	91350	26.04	33.34	55
2050	100755	140650	36.77	51.33	90

Source: India Stats

Water Requirements for Different Uses in India (1997-1998, 2010, 2025 and 2050)

Quantity in Billion Cubic Meter (BCM)

Uses	Year	Year – 2010			Year - 2025			Year - 2050		
	1997-98	Low	High	percent	Low	High	percent	Low	High	percent
Surface Water :										
Irrigation	318	330	339	48	325	366	43	375	463	39
Domestic	17	23	24	3	30	36	5	48	65	6
Industries	21	26	26	4	47	47	6	57	57	5
Power	7	14	15	2	25	26	3	50	56	5
Inland Navigation		7	7	1	10	10	1	15	15	1
Flood Control		-	-	0	-	-	0	-	-	0
Environment (1) Afforestation		-	-	0	-	-	0	-	-	0
Environment (2) Ecology		5	5	1	10	10	1	20	20	2
Evaporation Losses	36	42	42	6	50	50	6	76	76	6
Total	399	447	458	65	497	545	65	641	752	64
Ground Water :										
Irrigation	206	213	218	31	236	245	29	253	344	29

Domestic & Municipal	13	19	19	2	25	26	3	42	46	4
Industries	9	11	11	1	20	20	2	24	24	2
Power	2	4	4	1	6	7	1	13	14	1
Total	230	247	252	35	287	298	35	332	428	36
Grand Total	629	694	710	100	784	843	100	973	1180	100
Total Water Use :										
Irrigation	524	543	557	78	561	611	72	628	817	68
Domestic	30	42	43	6	55	62	7	90	111	9
Industries	30	37	37	5	67	67	8	81	81	7
Power	9	18	19	3	31	33	4	63	70	6
Inland Navigation	0	7	7	1	10	10	1	15	15	1
Flood Control	0	0	0	0	0	0	0	0	0	0
Environment (1) Afforestation	0	0	0	0	0	0	0	0	0	0
Environment (2) Ecology	0	5	5	1	10	10	1	20	20	2
Evaporation Losses	36	42	42	6	50	50	6	76	76	7
Total	629	694	710	100	784	843	100	973	1180	100

Source: India Stats

Growing Nation, Expanding Footprint

'The footprints you leave behind will influence others. There is no person who at some time, somewhere, somehow, does not lead another.'

—Unknown

Water—essential for all human activities, is also a key input for practically all economic activities like agriculture, industry, energy production and transport as well as for several domestic purposes. On an average, individuals require approximately 20 litres of water daily to meet their basic needs. However, there is a huge disparity in water availability and thereby usage, across the globe. Europeans use an average of 200 litres per day and North Americans 400 litres per day. Yet, over billion people have access to only about 5 litres per day. Even, this limited quantity of water is largely contaminated, which is causing major health problems across the world. It has been estimated that contaminated water causes around 1.8 million deaths per year. Production of essentials like food and other products and activities depend completely on the existence of a strong network of water provisioning. Given the massive dependence of India's population on agriculture and its deviating rain patterns, India's water policy framework needs immediate examination and reshaping. Costs, connected with all stages of water sourcing till its reach in end user's hands, are increasing. Then there is also the less visible cost—the time dedicated to collecting, hauling, storing, treating, and distributing water—which gets added to the cost of drought, desertification, and falling water tables. Stern N., Climate Change expert—London School of Economics[1] mentions that most developing countries which are facing water deficit lack sufficient storage options to manage annual demand. Moreover, he stated—'Inappropriate water pricing and subsidized electricity tariffs that encourage the excessive use of ground water for pumping also increase vulnerability to changing climatic conditions.' Unfortunately, for the globe at large and India in particular, the analytical and policy research contributions remain largely unconnected to the prevailing legal framework and the mechanics of international cooperation for adaptation or mitigation. There is complete absence of holistic perspective to water economics of the country.

Climate change is going to radically alter the scenario of water availability and usage, since water is the most climate-sensitive environmental resource. Already, water scarcity in various manifestations is exacerbating existing inequalities, contributing to mortality and morbidity[2]. The impact of climate change will be an additional trouble factor towards accentuation of people's water woes. It is amply observed that climate change and development are in complete tandem with each other as complementary concerns and not to be seen pursuing competing interests[3], while on the path of a sustainable growth trajectory.

Water Footprint: The Concept

The Water footprint of an individual, community or business is defined as the total volume of freshwater that is used to produce the goods and services that are consumed by each for their functioning. The water footprint shows human appropriation of the world's limited freshwater resources and thus provides a basis for assessing the impact of goods and services on freshwater systems and facilitates formulation of strategies to reduce the impact. The water footprint is a geographically explicit indicator that shows not only volumes of water used and pollution, but also the location and timing.

The water footprint is a new concept in the field of water management. It is expected that if water consumption or the need of a country is assessed on the basis of its water footprint, only then an efficient water management system and practice can be achieved. Human activities consume and pollute a lot of water. Though most of the water usage occurs in agricultural production, there are also substantial water volumes consumed and polluted in the industrial and domestic sectors. Water consumption and pollution can be associated with specific activities, such as irrigation, bathing, washing, cleaning, cooling and processing. Total water consumption and pollution are generally estimated as the sum of a multitude of independent water demanding and polluting activities. There has been little attention paid to the fact that, in totality, total water consumption and pollution relate to what and how much communities consume and to the structure of the global economy that supplies the various consumer goods and services.

The water footprint of a business consists of two components—

- The supply-chain water footprint—water used in the producer's supply chain (indirect water use)
- The operational water footprint—water used for producing/ manufacturing or for supporting activities by the producer (direct water use)

Water Audit plays an important role in measuring water footprint. Alongside monitoring of water usage, a water audit can identify specific measures to reduce the water footprint. These measures may include using water saving technologies, water conservation measures, and wastewater treatment. It is extremely important for businesses to know about its water consumption rate. Once a business has information about its water consumption, particularly about the water-intensive parts of the business, it can use this information for formulating an appropriate action plan, prioritizing short, medium and long term measures. This should also include analysis of proposals with timescales and return on investments, detailing how the business' water footprint can be reduced. Defining the boundaries of global water footprint for the businesses and measuring their total water usage including embedded water and water efficiency within global supply chains is going to become as important as measuring their energy use. And the first step towards attaining success in this is to begin with measuring the business water footprint accurately.

The concept of embedded water has become extremely important in the international debate on water and has been explained by John Anthony Allan of King's College. It has also helped people to understand their real water footprint, which measures the amount of water consumed by people on an average in the day, not only directly, but also virtually.

Businesses, because of the excess water scarcity and its effect on the business, have now started considering water risks. Water risks for businesses are categorized as below—

- Physical risks—lack of water availability, interruption of supply, deterioration in water quality
- Regulatory risks—more stringent targets, restrictions on supply, which are likely to lead to rise in prices of water supply and its treatment
- Reputational risks—competition for supply, food safety
- Financial risks—increased costs of water supply and water treatment and in entire supply chain
- Investment risks—increased risks to business and considerable spatial variation of these risks, have made it difficult for investors to gauge the risk and are affecting outcomes from the business decisions

In the present competitive times, knowledge of potential water scarcity hotspots across the supply chain as well as keeping abreast of potential technical or socioeconomic solutions will become increasingly important for any business strategy. From an investment perspective, it would be prudent during any acquisition to be aware of the degree of 'water richness', in the same way that certain industries have moved to countries with cheap labour. In addition, future investment potential within any existing company portfolio should consider water availability both as a key performance and key risk indicator. Water management, therefore, needs to be included as part of a wider water stewardship that addresses a more holistic approach related to a business's freedom to operate, as well as opportunities to increase market share through brand enhancement and competitiveness. Developing most accurate water footprint can help identify such crucial parameters. Water stewardship should not be tackled in isolation or with a tunnel vision. The linkages between water footprints, energy, carbon footprints and food production are key to future company management and investment strategies. There needs to be a strategic approach that addresses the whole water cycle along with carbon and energy requirements with respect to the entire supply chain.

Water Footprint of a Nation

The water footprint of national consumption is defined as the total amount of fresh water that is used to produce the goods and services consumed by the inhabitants of the nation. Consumption of agricultural products largely determines the global water footprint related to consumption, contributing 92 percent to the total water footprint. Consumption of industrial products and domestic water usage contribute 4.7 percent and 3.8 percent respectively. The global average water footprint related to consumption is 1385 m3/yr per capita over the period 1996-2005. The water footprint of Indian consumption was 987Gm3/yr in the period 1997-2001 i.e., 980m3/yr per capita[4]. However, while India contributes 17 percent to the global population, the people in India contribute 13 percent to the global water footprint. Average water footprint in US is 2842 m3/yr per capita and that of China is 1071 m3/yr per capita.

The four major direct factors determining the water footprint of a country are—

- Volume of consumption (related to the gross national income)
- Consumption pattern (e.g. high versus low meat consumption)
- Climate (growth conditions)
- Agricultural practice (water usage efficiency)

Water resources of the planet are being subjected to mounting pressures in the form of consumptive water use and pollution. Attempts have been made so far only to address issues related to freshwater availability, usage, and management at local, national, and river basin scale. The recognition that freshwater resources are subject to exploitation and globalization has led a number of researchers to argue for the importance of placing freshwater issues in the global context[5]. Freshwater withdrawals have tripled over the last 50 years. Demand for freshwater is increasing rapidly by 64 Billion Cubic Meter (BCM) a year (1 cubic meter = 1,000 liters). The world's population is growing roughly at the rate of 80 million people each year. Changing lifestyles and eating habits in the last three decades have led to the requirement of a higher amount of water consumption per capita[6]. Food and Agriculture Organization (FAO), has assessed the fresh water needs of the composition of meals and allowing for post-harvest losses, the present average food ingest of 2800 kcal/person/day may require roughly 1000 m^3 per year to be produced. With the world population at around 7 billion now, water needed to meet the essential food requirements would be around 7000 km^3[7]. It further states, most water used by agriculture stems from rainfall stored in the soil profile and only about 15 percent of water for crops is provided through irrigation. Irrigation, therefore, needs 900 km^3 of water per year for food crops (to which some more water must be added for non-food crops)[8].

Since the 1960s, the world food system has responded very ably to a doubling of the world population, providing more food per capita at progressively lower prices. Global nutrition has consistently improved. This performance was possible only due to a strategic system in place which combined high-yielding seeds, irrigation, plant nutrition and pest control. In the process, huge quantities of water were appropriated to agriculture. As population keeps growing, albeit at a decreasing rate, more food and livestock feed need to be produced in the future and more water is needed for this purpose[9]. Worldwide, agriculture accounts for 70 percent of all water consumption, in comparison with 20 percent for industrial usage and approximately 10 percent for domestic use.

With the growth rate of population estimated to be around 1.0 percent per annum, and the GDP growth predicted at around 6.8 percent per annum for the period 2005-30, the share of agriculture in GDP is expected to plunge from 19 percent to 10 percent[10]. But with mounting population coupled with an increase in wealth, the increasing caloric intake of the nation's population will be the undercurrent to be observed while evaluating the challenge of augmenting water resources. On the whole, unconstrained demand resulting from the production growth, driven by the rapid increase in demand for food and feed crops, particularly rice and wheat, would mean that in the year 2030 agriculture will account for almost 1,200 billion m^3 or 80 percent of total water demand—almost twice that of agriculture's water demand in 2005[11]. Further, it has been analysed that India faces a large gap between current supply and projected demand for water—which amounts to a whopping 50 percent of the entire demand made. This gap is caused due to a rapid increase in demand for water for agriculture, straddled with a limited supply of infrastructure[12]. India, however, also faces an aggregate gap of 50 percent across all basins, driven by very rapid growth in agricultural as well as municipal and domestic demand[13]. The projected municipal and domestic water demand will also double by 2030, to 108 billion m^3, while the projected demand from industry will quadruple to 196 billion m^3 (13 percent), pushing overall demand growth close to 3 percent per annum[14]. A major tentative factor that may affect the size of this gap is climate change and its consequences. Its most expressive effect is likely to be an accelerated melting of the Himalayan glaciers on which several of India's river systems are dependent, particularly the western rivers such as the Indus, which relies on snowmelt for approximately 45 percent of its water flow[15].

A study by the International Water Management on India's future water consumption reports that total water demand for the Business As Usual (BAU) scenario is projected to increase by 22 percent by 2025 and by 32 percent by 2050. The domestic and industrial sectors account for a substantial part of the additional water demand, 8 percent and 11 percent, respectively, of the total water demand by 2025, and 11 percent and 18 percent, respectively, by 2050. Moreover, the domestic and industrial sectors will explain for about 54 percent of the additional water demand by 2025, and for more than 85 percent by 2050[16]. The BAU scenario envisages significant water transfers from the irrigation sector to other sectors by 2050. The combination of higher irrigation efficiencies and large ground water irrigated areas decreases the demand for surface water irrigation between 2025 and 2050. While the total irrigation demand decreases by 38 bm^3, the demand for surface water irrigation is estimated to decrease by 46 bm^3[17]. Further, it also states that India's population is likely to reach 1,500 million by 2050, then grain needs, under well-fed scenarios, is projected to boost to around 494 million metric tons (MMT)—higher than a 150 percent increase from the current level[18]. This demand for food grain production will only be met if water is sufficiently supplied with technological supplements. In the base case projection for India, of the projected 685 million metric tons of food production in 2030, 175 million tons are expected to be rainfed, leaving 510 million tons of irrigated production. Fundamental to these numbers is the assumption that existing rain-fed lands would lessen to some extent as a result of some getting converted to irrigated lands, and that additional crop demand will primarily come from additional irrigated lands[19].

The popular movement to protect India's rivers can be fuelled by implementing the National Water Policy (2012) towards building a low-water economy, for which appropriate pricing of water becomes crucial. As regards pricing of water, there continues an ongoing debate whether water is a public/social utility or an economic good.

In rich countries, people generally consume more goods and services, which translate into increased water footprints. However, it is not consumption volume alone that determines the water demand of people. The composition of the consumption package is relevant too, because some goods in particular require a lot of water (bovine meat, rice etc). In many poor countries it is a combination of unfavourable climatic conditions (high evaporative demand) and bad agricultural practices (resulting in low water productivity) that contributes to a high water footprint. Some of the underlying factors that contribute to bad agricultural practices and thus high water footprints include improper water pricing, the presence of subsidies, the use of water inefficient technologies and lack of awareness of simple water saving measures among farmers.

Managing and Reducing Water Footprint

- Break the seemingly obvious link between economic growth and increased water usage, for instance by adopting production techniques that require less water per unit of product
- Water productivity needs to be increased by applying various advanced techniques such as rainwater harvesting and supplementary irrigation
- Shift to consumptions patterns that require less water, for instance by reducing meat consumption
- Water cost should be reflected in the pricing of goods and services

- Shift production from areas with low water-productivity to areas with high water productivity, thus increasing global water usage efficiency

Mandatory CSR Spend in the Companies Bill: Industry Views Divided

The 2 percent mandate on Company Social Responsibility (CSR) spend under the Companies Act 2013 incorporates for the first time mandatory CSR activities which includes water. In case corporates are not able to adhere to the mandated limit, they are supposed to at least declare in their annual report as to why its target of CSR activities has not been met.

Companies Act, 2013 (Companies Bill 2012) was passed in the parliament and made into a legislation in August 2013, therefore coming to be known as the Companies Act 2013. This is definitely a welcome change as it would go a long way in fostering sustainable and participative model of industrialization coupled with inclusive growth. Through this provision, companies' involvement with local community shall become more decentralized and less tenuous considering the industry had several key infrastructure and other industrial projects stalled due to local stakeholders' concerns and fears. Through this 2 percent provision, the Act also paves way for companies to be more innovative to the extent that it can leverage its market position to gain entry into previously untapped markets through impact investment. Similarly, social business will get a fresh lease of encouragement through the CSR provision. It has been seen that the Act does not define CSR, it provides a set of main and related activities that may form part of CSR. By doing so, the Act tries to provide flexibility to entities to choose its own area of charitable work. This clause, therefore, provides more autonomy than guidance to the corporate sector. Schedule VII of the Act among other activities covers reducing child mortality and improving maternal health, promoting gender equality and health; combating human immunodeficiency virus, acquired immune deficiency syndrome, malaria and other diseases; and ensuring environmental sustainability.

A closer look at the above activities would give us a clearer picture about the role that clean and hygienic water plays in sustaining human capabilities in rural and urban areas in India. One of the primary reasons for infant mortality is due to water borne diseases in India. It is estimated that about 2,000 diseases in India are caused due to consumption of poor or bad quality water. Access to clean drinking water improves public health conditions and also comes as a boon to women folk, especially those living in rural areas because majority of their time is spent in fetching water from faraway places. In such a situation, a substantial portion of their time which would have otherwise been invested in forming a part of the labour workforce goes wasted thus leading to more impoverishment. Access to water has manifold effects that are far-reaching and has crowding-in effect for several other social development indicators. The activities included in Clause 135 highlights the importance of water in achieving human development.

Corporate Initiatives to Reduce Water Footprint

Mahindra Rise[20]

The results of the study for 2011-12 were tabled by the Mahindras. It has identified opportunities to arrest wastage and contamination of water across sectors. In the reporting year, the absolute water consumption decreased by 12.74 percent from 7,255,728 m to 6,331,681 m as compared to the previous year. A 360° water conservation drive led to reduction in the specific water consumption at AD[21], Systech[22], MLDL[23], FSS[24] and MVML[25] as compared to previous year. New projects at Mahindra Intertrade, Kanhe and Nashik locations have resulted in the increase of specific water consumption this year compared to previous year. In harmony with the Rajiv Gandhi Mission for Watershed Management, the Farm Division initiated an integrated watershed development project in Damoh, Madhya Pradesh to drive environmental, economic and social development within the community. In the reporting year, Phase 1 of the project has been successfully completed empowering 20,000 people in 32 villages. The core purpose of watershed management is to ensure rainwater harvesting to provide the local community with year-long access to water. Higher yield of water would also translate to better avenues of irrigation, resulting in generation of higher income for the farmers. The watershed project has also been instrumental in generating employment for local labourers, helping them earn anywhere from INR 200 to INR 300 per day, and thereby controlling migration. Similar success was chalked up in water consumption with a 13.45 percent reduction in three years against targets of 2 percent in three years and 5 percent in five. Micro-irrigation offers tremendous benefits to the farmer which includes over 25 percent water savings, reduced expenditure on labour and fertilizer and higher productivity. By virtue of this development, Mahindra will be able to help the farmer to better utilize scarce water resources and thereby contribute to overall water conservation in the country[26] A comprehensive water foot-printing exercise has been done to understand and optimize our[27] water consumption[28]. The findings from this exercise will help the group to design more water-sensitive projects in the future.

Indian Hotels Company Limited (IHCL)[29]

Environment Awareness and Renewal programme at Taj Hotels is the cornerstone from where the company-wide movement of environment management initiates. It is a coordinated effort by IHCL to protect, conserve and restore the natural environment. It evolves from understanding each location's impact on the environment to adopting appropriate steps for protecting and conserving it. Across hotels, IHCL have started several good practices and projects aimed at tapping solar and wind energy, installing water recycling plants, generating biogas from organic waste and conserving energy. Many of IHCL hotels have attained zero waste water discharge and have sustained partnerships for responsible recycling of waste[30]. If you take a stroll on any beach, chances are you will come across plastic water bottles scattered all around. It's not just an unpleasant sight but it's an environmental hazard as plastic takes thousands of years to decompose. So to prevent pollution caused by these bottles, IHCL hotels in Maldives have installed a unique distillation and bottling plant that converts sea water into potable water. This water is bottled in attractive custom-made glass bottles for guest consumption. This initiative has helped reduce the hotel's dependency on packaged drinking water and also reduce the garbage generation at the hotels.

Hindustan Construction Co. Ltd (HCC)[31]

HCC has begun recycling water and now uses 30 percent less water than before at construction sites. HCC has gained a lot in this process—despite the 2009 drought that hit Andhra Pradesh, which put a stop to the construction activity around the state, the company, however, was able to continue construction because it had been recycling water. As the first Indian company to endorse the United Nations Global Compact's CEO Water Mandate, HCC made it a point to embed the principles of water resource management in its works in the field of water. As a responsible corporate steward, HCC has also motivated other companies to join this initiative by presenting best practices of water stewardship at HCC at various local, regional, national, and international conferences and symposia[32]. Water meters have been installed on all HCC operating sites. HCC's total water consumption for the Financial Year 2010-11 stood at 1.78 million m^3 including ground water, surface water, rainwater, water from municipalities and water tankers. While, HCC recycled 3,74,372 m^3 water, which is 21 percent of all water consumed, across all its project sites.

Tata Chemicals[33]

The drivers of sustainability for the company are: reduce specific energy and water consumption to levels comparable to global benchmarks and be in the top ten of global best performers. Attain overall water neutrality and reduce/eliminate ground water usage especially from shallow aquifers which can affect the ground water table in the surrounding area. Conduct water resource management studies at all locations involving water intensive operations. Reduce water and energy consumption in usage of products through product design and promotion of better usage practices. Water conservation, reuse and recycling have been the key for Tata Chemicals' survival over the years. Manufacturing processes depend on Arabian Sea water, ground water (wells), surface water (rainfed lakes) and PHE supply water. All manufacturing processes are part of highly integrated inorganic chemical complexes, which facilitate conservation as a main environment management tool as seen in the Haldia and Mithapur plants which have successfully developed reuse / recycle of solid wastes from its operations. With the use of innovative seawater desalination technologies at Mithapur the use of ground water has been completely stopped. The Babraia plant has achieved benchmark levels in specific water consumption in the industry. Tata Chemicals believes that material conservation, energy and water conservation are also of prime importance for future sustainability. Mithapur has achieved 95 percent reduction in dependence on ground water in the last five years by adopting various in-house water conservation measures as well as management of the rainfed lake. Though urea-manufacturing process wholly depends on ground water, specific water consumption at the company is a benchmark amongst the nitrogenous fertilizers' manufacturers. Periodic water level and quality measurements are a part of the quality assurance system of Babrala. Babrala and Haldia fertilizer sections are zero discharge plants and mainly depend on recycled water, while chemical plants are major consumers of raw water. In the area of waste water recycling and reuse, Tata chemicals has pioneered recycling of water by integrating various manufacturing operations; recycle of condensate as boiler feed water at Mithapur.

Other Initiatives

Regulatory authorities like the Securities and Exchange Board of India (SEBI) may also consider initiating norms that will make it obligatory for companies to include disclosures of water usage in the supply chain, introduce water accounting, propose and plan a strategic module on future business development and its dependence on water in the supply chain and inclusion of water governance in company's project planning management. SEBI may also promote the concept of Water Impact Index and should rank organizations as per that Index. Responsible investors who believe in sustainable development may apply their mind and judgement based on the impact index before investing in the entity.

Further, credit rating agencies like CRISIL (Credit Rating and Information Services of India Ltd) launched the ESG India Index on January 29, 2008. It has yet to become a preferred benchmark amongst the investing community as the appetite for such products is relatively less especially during the turbulent times. It is seen, that by and large, asset managers in India are aware of the ESG risks and their impact on the financial performance of a company. However, they still do not integrate them into their investment practices for various reasons, one of them being that their own performance is linked to the performance of the asset they are managing where the yardstick is daily/monthly/quarterly performance, however, ESG risks usually play out in the medium to long term. Making money through stockmarket is not even one generation old in India. So, the focus of the investor is on just making money and they are indifferent to how the money is made. Therefore, investors are still not sensitive to how the company is making money in the short term (say by depleting the water resources of a region) and they are fine so long as the financial results of the company are good.

Conclusion

Though agriculture consumes the maximum share of the dwindling water resources, industry, too, in absolute numbers consumes a lot of water. It has to be careful in its water usage. In recent periods, visionary corporates have started performing responsibly with respect to water usage.

Global Reporting Initiative (GRI) has been guiding corporates how best to report their Sustainability Track Record of select parameters like energy and water. Dr Aditi Haldar, Head GRI India Chapter, describes brilliantly the nuances of such an important initiative—

'*GRI works towards a sustainable global economy by providing organizational reporting guidance. A sustainable global economy should combine long term profitability with social justice and environmental care. This means that for organizations, sustainability covers the key areas of economic, environmental, social and governance performance.*

A sustainability report enables companies and organizations to report sustainability information in a way that is similar to financial reporting. Systematic sustainability reporting gives comparable data, with agreed disclosures and metrics. GRI provides all companies and organizations with a comprehensive sustainability reporting framework that is widely used around the world. In the Guidelines there are indicators such as total water withdrawal by source; water sources significantly affected by withdrawal of water; and percentage and total volume of water recycled and reused.

Reporting the total volume of water withdrawn by source contributes to an understanding of the overall scale of potential impacts and risks associated with the reporting organization's water use. The total volume withdrawn provides an indication of the organization's relative size and importance as a user of water, and provides a baseline figure for other calculations relating to efficiency and use. The systematic effort to monitor and improve the efficient use of water in the reporting organization is directly linked to water consumption costs. Total water use can also indicate the level of risk posed by disruptions to water supplies or increases in the cost of water. Clean freshwater is becoming increasingly scarce, and can impact production processes that rely on large volumes of water. In regions where water sources are highly restricted, the organization's water consumption patterns can also influence relations with other stakeholders.

Likewise, withdrawals from a water system can affect the environment by lowering the water table, reducing volume of water available for use, or otherwise altering the ability of an ecosystem to perform its functions. Such changes have wider impacts on the quality of life in the area, including economic and social consequences. This Indicator measures the scale of impacts associated with the organization's water use. In terms of relations with other users of the same water sources, this indicator also enables an assessment of specific areas of risk or improvement, as well as the stability of the organization's own water sources.

The rate of water reuse and recycling can be a measure of efficiency and can demonstrate the success of the organization in reducing total water withdrawals and discharges. Increased reuse and recycling can result in a reduction of water consumption, treatment, and disposal costs. The reduction of water consumption through reuse and recycling can also contribute to local, national, or regional goals for managing water supplies.

While we see many large corporations like Tata Chemicals, SAIL, Aditya Birla Group, NMDC, ACC, Wipro, ITC, Coal India, Mahindra & Mahindra, Bajaj, GSK have shown initiatives through various aspects of water management, however the greater mass of India Inc has to demonstrate their efforts in a systematic manner. The measurement, management, change and communication are an integral part of these initiatives which every company can take advantage of. The reporting provides the opportunity to reflect, review on their strategy not only for their own operation but for the existence and continuance of our whole ecosystem including human beings. Reporting on the aspects of water although is one of the most common parameter in most of the companies preparing their sustainability reports, however the intent of the indicators are most often not imbibed. This is where the challenge lies for all of us – reporting cannot be a tick box disclosure. Reporting should provide the credible and trustworthy information backed by a standard framework and its true intent.

Beyond all this, the biggest question which is looming at us in this country where population is growing annually, is whether water is being considered and understood by all as a resource which is renewable yet not inexhaustible – it cannot withstand constant over-extraction and being depleted faster than being renewed. The bitter truth that we all need to face is that we do not have any substitute for water.'

The Pool is getting Murkier

'Destruction is a man's will... Nevertheless Prevention is also a man's will. It's a man's choice to choose between Destruction and Prevention.'

—Babu Rajan

India is richly endowed with water, with the nation's coastline stretching across 7,500 kilometres. The country is also well endowed with a network of rivers and blessed with the snow-capped mountains of the Himalayas.

Indian rivers have been classified into fourteen major, 44 medium and 55 minor and desert river basins. The major river basins cover 83 percent of the total drainage basin and contribute to 85 percent of the total surface flow, whereas medium and minor river basins share 8 percent and 7 percent respectively of the total surface flow. India also receives an average of 4,000 billion cubic meter (BCM) of rainfall every year. However, only 48 percent of the rainfall ends up getting drained into India's rivers and of which only a mere 18 percent is utilized due to a lack of storage facilities and related infrastructure.

What's more, India is not only failing when it comes to water storage, making matters worse is the growing threat of water pollution , a dynamic threat that has only added even more urgency to the nation's water crisis. Water pollution has emerged as one of the gravest environmental issue in India, which is also the major contributor to water scarcity. This gets aggravated by the pressure of increasing population, improper growth of industries, haphazard urbanization, energy intensive lifestyle and lack of environmental awareness, which are some of the impacts of the demographic changes on the environment.

Aggravating the situation are institutional deficiencies like inadequate infrastructure, lack of implementation of environmental rules and regulations, and insufficient enforcement of policies that has created a carefree and leashless attitude among water users. Practices like discharging of untreated effluent from industries and municipalities, use of non-biodegradable pesticides/fungicides/herbicides/insecticides, use of chemical fertilizers instead of organic manures—these environmentally damaging practices are common and widespread, and happen with little retribution.

The result has been—rising pollution of India's already precious and scarce water supply. Notably damaging is the pollution taking place in India's limited fresh water supply bestowed by nature, making it non-potable. Pollution causing huge adverse impact on India's freshwater resources is evident from the fact that 14 percent of the rivers in India are severely polluted, while another 19 percent are considered moderately polluted.

In India, the discharge of untreated sewage is the main cause for pollution for both surface as well as ground water. Household borne effluents is a key factor of pollution. Nearly 12.5 million people in India or 18.5 percent of households, do not have access to a drainage network, while only about 27 million i.e., 39.8 percent households are connected to open drains. In respect to underground sewerage, the availability is 30 percent and 15 percent in notified and non-notified slums, respectively.

Making matters worse—a majority of the government-owned sewage treatment plants remain shut most of the time due to improper design or poor maintenance or lack of reliable electricity supply to operate them. These plants also face problems like absentee employees and poor management. Most of the waste water either directly or indirectly affects the quality of water resources and makes them unfit for consumption.

Almost one-fifth of the centrally-funded sewage treatment plants in the country are 'non-operational', leading to millions of liters of untreated water either seeping into the ground as potential pollutant of ground water or being discharged into natural drainage systems and rivers every day. Besides, performance of another nearly one-fifth of the 152 sewage treatment plants (STPs) was found to be unsatisfactory.

The above findings are part of a report of the Central Pollution Control Board (CPCB)[1], which conducted performance evaluation of STPs. The report said—'Out of the 152 STPs, nine plants are under construction, 30 are non-operational, and performances of 28 plants are not satisfactory. A significant volume of waste water is not subjected to any treatment and is ultimately discharged into surface water bodies leading to deterioration of water quality.'

Ground water

Ground water is the major source of drinking water in both urban and rural India. It is also an important source of water for the agricultural and the industrial sectors. India possesses about 432 BCM of ground water that is replenished yearly from rain and river drainage, but only 395 BCM of this becomes useable. Of the 395 BCM, 82 percent goes for irrigation and agricultural purposes, while the remaining 18 percent gets divided between domestic and industrial sectors. Total static ground water available is approximately 10,812 BCM which is a huge resource that can be utilized.

The actual problem with ground water extraction is from the deeper depths (deep aquifers), where the rate of recharge decreases with the depth. In order to fulfil the enormous demand that even the public water infrastructure is unable to meet, ground water is being increasingly pumped from lower levels at much faster rates than what rainfall is able to replenish. In regions where rainfall is abundant, water is also available at lesser depths (shallow aquifers). Both these sources of aquifers are equally vulnerable to depletion—the former by virtue of the recharge rate and the latter because of easy access. The Delhi Jal Board, which is responsible for supplying potable water in the Capital city, estimates that water tables are dipping by an average of 0.4 meters a year.

During the past two decades, the water level in several parts of the country has been falling rapidly due to increase in extraction. The number of wells drilled for irrigation of both food and cash crops have rapidly and indiscriminately increased. India's rapidly rising population and changing lifestyles have also increased domestic requirement of water.

Water requirement for the industry is also showing an overall increase. There is an intense competition among users—agriculture, industry, and domestic sectors—which are driving the ground water table still lower. The quality of ground water is getting severely affected because of the widespread pollution of surface water. Besides the discharge of untreated waste water through bores, leachate (leached out water released from solid waste dumping sites. The action of moisture, temperature and pressure acting within a solid waste dump, releases a thick black watery substance which is called leachate), contaminate the ground water. The most unspoken source of ground water contamination is the contribution made by domestic users through individual bore holes.

Unchecked Depletion and Contamination

There has been a lack of adequate attention to water conservation, efficiency in water use, water reuse, ground water recharge and ecosystem sustainability. An uncontrolled use of the borewell technology has led to the extraction of ground water at such a high rate that often recharge is not sufficient. The causes of low water availability in many regions are also directly linked to the declining forest cover and soil degradation.

Contamination of ground water resources has become a major concern today. The solid, liquid and the gaseous waste that is generated, if not treated properly, results in pollution of the environment. This affects ground water due to the hydraulic connectivity in the hydrological cycle.

Water extraction without proper recharge and leaching of pollutants from pesticides and fertilizers into the aquifers has polluted ground water supplies. In addition, leachates from agriculture, industrial waste and the municipal solid waste have also polluted both surface water and ground water. Technically, a bore-well has to have a casing along the length of the bore hole, which has a cost associated. In order to save this expenditure, casing is installed only to a shorter depth of soft strata. Without casing, the bore hole is exposed to the surrounding strata, thereby exposing the percolated contaminations in the ground to reach further depths and contaminate the large ground water resource.

Pollution both in rural areas which leads to water depletion and industrial pollution faced in urban areas need to be addressed. Another indirect cause of water pollution is air pollution, as it leads to acid rains which pollute rivers, lakes and ponds.

Adding urgency to all of the above is that the framework and institutional architecture of water pollution, water use and the likes have yet to be developed. There is a need to pursue a multipronged response to an issue that has many angles to it. Mechanisms to meet the challenges of adaptation will need out of the box thinking at the national level. A combination of financial instruments has to be resorted to, for regulating carbon emissions and commodities professing these emissions. A judicious exploration of options like carbon taxes and permits, levies and auctions for carbon trading, ending fossil fuel subsidies and tax breaks, as well as taxes and transfers related to the consumption of goods that generate greenhouse gases or sequester carbon are needed, with necessary caution. There should be additional financing made available for the rural areas and its major source of occupation viz., agriculture; it should be based on estimates of the cost of mitigation; it should be reliable, consistent, and readily forthcoming.

The livelihood opportunity costs and consequences faced at the sectoral (agriculture, manufacturing, services and industries) level need to be understood through an ethnic and regional lens. This exercise should estimate India's diversity and gauge the segmentation in the kinds and spheres of pollution caused. Particular attention should be paid to methodologies that modify existing prices by taking into account such crucial contributions. Estimates should be revised regularly, whenever new awareness is forthcoming about the impact and consequences of climate change along with the incidence of costs and benefits.

Further, demographic pressures from population growth, age distribution, urbanization and migration create some of the greatest pressures on water resources while adversely affecting their quantity and quality. These demographic strains directly affect water availability and quality through increased water demands, consumption and through pollution resulting from water usage[2]. Pollution, typically, refers to chemicals or other substances in concentrations greater than what would occur under natural conditions. Some of the major forms of pollution include fecal coli forms, industrial organic substances, acidifying substances from mining aquifers, atmospheric emissions, heavy metals from industry, ammonia, nitrate and phosphate pollution from agriculture, pesticide residues (again from agriculture), sediments from human-induced erosion in rivers, lakes and reservoirs and salinization[3].

Water pollutants are categorized as point or non-point according to their primary sources. Point sources (PS) are pollutants from pipeline discharges and other readily identifiable sources. Non-point sources (NPS) are pollutants mobilized by precipitation as it flows over land which later infiltrates the soil. The non-point source load pollutant in a drainage basin is generally functions of precipitation patterns which convey the range of human activities occurring in the basin (especially agriculture).

It has been noted that the household generated effluents contribute a substantial proportion to water pollution occurring in the country. Untreated effluents from households pollute both surface and ground water sources. Local governments (city corporations, municipalities and panchayats) have the responsibility of water supply and sanitation and are supposed to treat the effluents as per the national water pollution standards or minimal national standards (MINAS)[4]. However, an alarming 70 percent of the effluents go untreated and are disposed off into the environment. Further, of the total effluent treatment capacity of 11,554 MLD in the country, about 70 percent (8040 MLD) is created in 35 metropolitan cities. Metropolitan cities treat approximately 52 percent of their wastewater. Delhi and Mumbai account for around 69 percent of the treatment capacity of metropolitan cities. This indicates that smaller towns and cities have very little wastewater treatment capacity[5]. The process of urbanization is accompanied by the transformation of natural land surfaces into impervious surfaces, such as streets, parking lots, roofs and other types of structures that block the percolation of rainwater and snowmelt into soil. The negative externality is that such construction increases the flow velocity of water over the land surface, carrying polluting materials into the receiving water systems, degrading water quality and causing local pollution problems[6].

Nitrate pollution of ground water is emerging as a foremost environmental issue in developing as well as industrialized countries[7]. In India, ground water resources in vast proportions are contaminated with fluoride and arsenic. Fluoride problems exist in 150 districts in 17 states in the country with figures indicating Orissa and Rajasthan being the most severely affected. High concentration of fluoride in drinking water causes fluorosis and ailments like weak bones, weak teeth, and anemia[8].

Regions of the world that are sustained by agriculture are more prone to river depletion because of irrigation. Research reports indicate agriculture to be the main source of nitrate pollution of ground water and surface water, as well as a principal cause of ammonia pollution. It is also a major contributor to the phosphate pollution of lakes and waterways and to the release of methane and nitrous oxide into the atmosphere. The improper use of pesticides has had a considerable negative impact on the environment, resources and consequently human health[9]. In arid areas, the return flow from agriculture itself and multiple reuses of water have resulted in rapid degradation in quality, in particular, combined problems of water pollution and water scarcity[10]. In areas defined as development access areas and previously where forests existed we can see that extensive tree-cutting has led to an increase in natural water run-off, lack of terrain reforestation and accelerated soil erosion creating more sediments, and lack of run-off control in steep terrain areas.

Industrial Pollution: Major Concerns

As per existing standards, industries are supposed to treat their effluent to the prescribed levels before discharging into the natural water bodies (rivers, streams or sea). In reality, however, many industries discharge the effluent into the natural water bodies directly and untreated. In many places, the effluent is discharged into the sewer lines, intended to be connected to the common or municipal treatment plants, instead is discharged untreated into the natural water bodies. On the other hand, there are industries that follow the standards consciously and continuously strive to improve, but not many of the industries are following these rules and regulations.

Direct discharge not only contaminates the water but also sediments downstream because of the industrial discharge. Water that has a high organic content (called biochemical oxygen demand, or BOD) often appears cloudy or foamy, and is characterized by the rapid growth of algae, bacteria and slime. The growths of these organisms deplete the level of oxygen in the water. It becomes almost impossible for aquatic beings like fish, insects, amphibians and many species of aquatic plants to live and breed in such oxygen-depleted water. If the water discharged still maintains a high temperature, this 'thermal pollution' may also affect the aquatic ecosystems downstream, which as a result has to adjust to a temperature that is higher than normal[11]. The toxicity levels and lack of oxygen in the water can damage or completely destroy the aquatic ecosystems downstream as well as lakes and dams, ultimately adversely affecting estuaries and marine coastal environments. In international river basins, routine pollution and polluting incidents such as industrial accidents and spillages may have transboundary effects. Significant pollution sources in river basins, such as large industrial plants, may be termed 'hot spots' and prioritized for cleanup within a river basin management plan[12].

Of the sources of discharge discussed earlier, point sources (PS) are easy to control and more direct and quantifiable, and in many developed countries its mitigation has been linked to treatment, achieving lower contaminant concentrations before discharge. Non-Point Source (NPS) pollution occurs when contaminants from diverse and widely spread sources are transported by run-off into rivers, lakes, wetlands, ground water and coastal areas. This type of pollution is more difficult to address as there are a large number of sources, for example, varied agricultural areas all of which are using pesticides and nutrients. Today, however, NPS pollution is receiving more attention as its impact is becoming evident over large areas in lakes, streams and ground water and can also be linked to the degradation of aquatic freshwater and marine ecosystems[13].

In addition to the impact on the aquatic environment through the discharge of cooling water, fossil fuel (especially coal based) power generation is also responsible for air pollution. Emissions of oxides of sulphur and nitrogen are responsible for acid rain, the deposition of which results in the degradation of ecosystems, as well as damage to agricultural production and to buildings. As a result of these measures, the use of low sulphur coal, the deployment of dust filters, flue (flue-gas desulfurization (FGD) is a set of technologies used to remove sulfur dioxide (SO2) from exhaust flue gases of fossil fuel power plants, and from the emissions of other sulfur oxide emitting processes) gas desulphurization and nitrogen oxide control technologies are common practices at modern fossil fuel generating plants. Atmospheric emissions from coal combustion are now the major challenge confronting continued deployment and development of coal-fuelled power plants.

Overall Water Quality in India[14]

The water quality monitoring results obtained during 1995 to 2009 indicate that the organic and bacterial contamination continue to be critical in water bodies. This is mainly due to discharge of domestic waste water, mostly in untreated form, from the urban centres of the country. The municipal corporations are unable or unlikely to treat the increasing load of municipal sewage flowing into water bodies. Secondly, the receiving water bodies also do not have adequate water for dilution. Therefore, the oxygen demand by organisms in the rivers and ponds and bacterial pollution is increasing every day. The water quality monitoring results were analysed with respect to the indicator of oxygen consuming substances BOD and indicator of pathogenic bacteria (total coliform and faecal coliform). The result of such analysis shows that there is gradual degradation in water quality, with an increase in the density of BOD and Coliform during 1995 to 2009[15].

Status of Water Quality in India (2009), states that there is a gradual, yet consistent degradation in water quality. However, in rural areas the existing water purification process, under the rural water supply schemes, involves accessing water from surface (tank, river, canal, etc.) sources; processing it through slow sand filters (SSFs), chlorination and pumping to an overhead tank for subsequent piped supply to the community or households. In the case of fluoride affected areas, defluoridation plants are established at the community or village level. In some areas, domestic defluoridation filters were provided to the poor households. Despite a number of initiatives towards providing safe drinking water in rural areas, water related health problems seem to persist in a significant way. Further, it is increasingly agreed in development circles that water shortages and increasing pollution are to a large extent socially and politically induced challenges inherently built into the societal structure, which indicates that there are issues which can be addressed by alterations in the demand for water and its use through increased awareness and ultimately education and water policy reforms[16].

Most importantly, the legal rules and regulations have to be updated. In the present scenario of increasing ecological crisis due to climate change, the whole question of laws relating to water including pollution needs to be fundamentally examined. Due to water being a state subject, there is no comprehensive water law in India at the central level. However, at the federal level, the environment law framework in the country addresses issues with regard to water pollution and environmental standards. The Water (Prevention and Control of Pollution) Act, 1974, The Water (Prevention and Control of Pollution) Cess Act, 1977, and the Environment Protection Act, 1986 together constitute legislative and regulatory framework to address water allocation and usage from the pollution standpoint alone. Therefore, in the absence of comprehensive water legislation, there must be more efforts to assimilate the knowledge of water conservation and efficient use so as to enable policymakers and implementing agencies to understand the challenges and strategies involved in adopting and implementing necessary measures for managing water in a sustainable manner.

The Policy (T)angle

'Wishful thinking is not sound public policy.'

—Bjorn Lomborg

In most parts of the world—policymakers and business leaders face a great challenge in the form of water insecurity. The subject has traditionally been dominant as the topic of debate amongst public-sector leaders and non-governmental organizations (NGO). But now, increasingly, policymakers recognize that allocation strategies, pricing and advanced technologies are essential for effectively managing scarce water supplies. The private sector and business leaders are making efforts to understand the difference between effective and ineffective policies, and in deploying contemporary technologies for upscaling and upgrading the infrastructure.

Considerably greater attention is now being paid, by the companies in the private sector, to the water ecosystem, in which they operate. As these companies continue to advance to the next level of water and related technologies, their role in generation of new age policies is becoming increasingly relevant. Taking the cue, policy leaders have started to acknowledge, the private sector's influence and involvement in drawing up empowering guidelines. This engagement, however, is not mainstream, and many more, from both the groups need to join the dialogue and engage with other stakeholders too.

The nation's water policy, therefore, is an extremely important and authorized institutional device, as it embodies the government's objectives, direction, and plans with regard to management of water resources. In keeping with the democratic framework and growing importance of water in the society, it is becoming increasingly relevant, that this process is accomplished, in consultation with other experts and the general public.

The availability of fresh water is already under heavy strain and the trend is likely to continue with the expanding population and demands of a growing nation. In combination with the forecasts of impact due to climate change, further deterioration and a conflict-like situation is portended. In spite of this alarming state, the consciousness about the issues around water's efficient use and proper management, the double impact of pollution continue to be relatively low.

The Present Framework

With the aforesaid scenario as the backdrop, the recently drafted National Water Policy 2012[1] is an effort towards developing a framework that addresses some of these concerns that may include—

- Large parts of India have already become water stressed. Rapid growth in demand for water due to population growth, urbanization and changing lifestyle pose serious challenges to water security
- Issues related to water governance have not been addressed adequately. Mismanagement of water resources has led to a critical situation in many parts of the country
- There is wide temporal and spatial variation in availability of water, which may increase substantially due to a combination of climate change, causing deepening of the water crisis and incidences of water related disasters, i.e., floods, increased erosion, and increased frequency of droughts, etc

- Climate change may also increase the sea levels. This may lead to saline intrusion in ground water aquifers, surface waters, and increased coastal inundation, adversely impacting habitations, agriculture, and industry in such regions
- Access to safe water for drinking and other domestic needs still continues to be a problem in many areas. Skewed availability of water between different regions and different people in the same region, also the intermittent and unreliable water supply system, has the potential of causing social unrest
- Ground water, though part of hydrological cycle and a community resource, is still perceived as an individual property, and is exploited inequitably, and without any consideration to its sustainability, leading to its over-exploitation in several areas
- Water resources projects, though multidisciplinary with multiple stakeholders, are being planned and implemented in a fragmented manner, without giving due consideration to optimum utilization, environment sustainability, and holistic benefit to the people. Inter-regional, inter-state, intra-state, as also inter-sectoral disputes in sharing of water, strain relationships and hamper the optimal utilization of water through scientific planning on basin/sub-basin basis
- Grossly inadequate maintenance of existing irrigation infrastructure has resulted in wastage and under-utilization of available resources. There is a widening gap between irrigation potential created and utilized
- Natural water bodies and drainage channels are being encroached upon, and diverted for other purposes. Ground water recharge zones are often blocked
- Growing pollution of water sources, especially through industrial effluents, is affecting the availability of safe water, besides causing environmental and health hazards. In many parts of the country, large stretches of rivers are both heavily

polluted and devoid of flows to support aquatic ecology, cultural needs and aesthetics

- Access to water for sanitation and hygiene is an even more serious problem. Inadequate sanitation and lack of sewage treatment are polluting the water sources
- Low public consciousness about the overall scarcity and economic value of water results in its wastage and inefficient use
- The lack of adequate trained personnel for scientific planning, utilizing modern techniques and analytical capabilities incorporating information technology constrains good water management
- A holistic and inter-disciplinary approach towards water related problems is missing
- The public agencies in charge of taking water related decisions tend to take these on their own without consultation with stakeholders, often resulting in poor and unreliable service characterized by inequities of various kinds
- Characteristics of catchment areas of streams, rivers and recharge zones of aquifers are changing as a consequence of land use and land cover changes, affecting water resource availability and quality

The policy also upholds that water resources need to be governed by certain basic principles, in order to have some commonality in approaches in dealing with planning, development and management of water resources. These basic principles as per the draft are—

- Planning, development and management of water resources need to be governed by common integrated perspective considering local, regional, state and national context, having an environmentally sound basis, keeping in view the human, social and economic needs

- Principle of equity and social justice must inform use and allocation of water
- Good governance through transparent informed decision making is crucial to the objectives of equity, social justice and sustainability. Meaningful intensive participation, transparency and accountability should guide decision-making and regulation of water resources
- Water needs to be managed as a common pool community resource held, by the state, under public trust doctrine to achieve food security, support livelihood, and ensure equitable and sustainable development for all
- Water is essential for sustenance of ecosystem, and therefore, minimum ecological needs should be given due consideration
- Water, after meeting the pre-emptive needs for safe drinking water, sanitation and high priority allocation for other domestic needs (including needs of animals), achieving food security, supporting sustenance agriculture and minimum ecosystem needs, may be treated as economic good so as to promote its conservation and efficient use
- All the elements of the water cycle, i.e., evapotranspiration, precipitation, run-off, river, lakes, soil moisture, and ground water, sea, etc., are interdependent and the basic hydrological unit is the river basin, which should be considered as the basic hydrological unit for planning
- Given the limits on enhancing the availability of utilizable water resources and increased variability in supplies due to climate change, meeting the future needs will depend more on demand management, and hence, this needs to be given priority, especially through (a) evolving an agricultural system which economizes on water use and maximizes value from water, and (b) bringing in maximum efficiency in use of water and avoiding wastages

- Water quality and quantity are interlinked and need to be managed in an integrated manner, consistent with broader environmental management approaches, inter alia, including the use of economic incentives and penalties to reduce pollution and wastage
- The impact of climate change on water resources availability must be factored into water management related decisions. Water using activities need to be regulated keeping in mind the local geo climatic and hydrological situation

At a Quick Glance

As the present draft National Water Policy varies significantly, both in terms of scope, and approach to management of water resource from its forerunners[2], the importance of a transparent consultation process was recognized and some efforts have been made before proceeding forward. In the absence of a wider participation of communities in this process, the sensitive issues around availability and distribution of water could lead to gross discontent. Although some critics and stakeholders would have preferred an even wider and more open consultation, the draft was publicized well enough and relevant delegates were invited for their inputs. It has been a reasonable effort which could have taken a slightly more democratic approach. The consultations with several water experts and NGOs had been conducted for over two years. In the same period, there could have been more involvement of the Panchayati Raj institutions which would have ensured that voices of communities directly impacted by climate change were heard, which in turn would have aided successful implementation of the policy.

An overview makes it seem as if the policy has taken an all-encompassing approach—priority for safe and clean drinking water and sanitation; incentivizing water recycling; stress on efficient water usage across agriculture, industry and domestic sectors; improvement in waste water treatment and rural water supply. The present water policy introduces several new measures and takes a different route in addressing the water challenge approach, from the previous ones. But, in the absence of clear guidelines and legally enforceable mechanisms, it is as prone to being ambiguous and largely ineffectual as the previous ones.

To begin with, a statement such as 'water for such human needs should have a pre-emptive priority over all other uses' needs to be qualified in a much better way. India has voted affirmatively at the United Nations General Assembly Resolution on the Right to Water in 2010[3]. Therefore, for the fundamental rights to water to have more meaning, the policy needs to be backed up by well-defined parameters that demand precise service standards. Some lessons can be learnt from the limitations of the previous policies that had stressed upon ecosystem needs, and specified that minimum flows will be maintained in rivers. Assessing the damage to the rivers that have turned into sewers or the wetlands that have been encroached upon, it is evident that mere statement has had no impact in the absence of strict guidelines and monitoring authority. Further, as we will discuss shortly, the implementation of the part of the policy that aims at ascertaining basic access to water while establishing economic value and full cost recovery is a daring intention.

The policy has a clear statement for climate change, and states that distinctive consideration towards micro level alleviation will be by augmenting the capabilities of communities to transition to climate resilient technological alternatives. Going beyond the statement what needs to be discussed and assessed is the state of present water infrastructure and its ability to transform in the face of daunting climate change challenges. The country can also learn from some of the global sustainable practices that aid in increasing soil's capacity to retain moisture, organic farming et al. While a big thrust is required on adding and refurbishing infrastructure, time tested strategies on flood forecasting and regulation can be deployed without further ado.

In one of the sections the draft talks about the development of water resources projects including large hydro power projects with diverse objectives—all with proper provision for storage. In general, it may be recommended that at the planning stage when available water resources are taken into account the impact of climate change on its future availability should also be considered. Effective ways to store precious water, should however, extend beyond new projects being planned and should be seriously pursued in existing facilities with equal rigour. Conservation of water is not just a duty of the state but also a responsibility of its citizens. As has been said earlier, the traditional and low cost solutions to storing water aided by adequate investments towards technical enhancements need to be encouraged at all level.

An important area where the new draft water policy proposes significant transformation is to recommend new institutional apparatuses to strengthen the economic value of water. As a major departure from the predecessors, the policy hints at a reduced role of the state in public services. The latter may continue to act as a strong facilitator and a regulator, while the local communities and private sector work towards newer technologies and other solutions for better delivery. The new policy includes some welcome change in priorities for water to be treated as a community resource under state ownership. These public-private partnerships engaging with the communities and other stakeholders may be the framework the country needs going forward. There are skeptical views about privatization, but the state at which the present infrastructure is and the investment needed to upgrade and upscale, this model is definitely worth giving a chance. The establishment of Water Resources Regulatory Authorities (WRRA) in all the states after experimentation in Maharashtra could be one of the ways to regulate the sector, but in the absence of an impressive track record with the pilot or reform in its structure, it may prove to be regressive. While the regulatory authority to supervise the partnerships and overall distribution is required, it should have better representation of stakeholders in each state and should clearly define its role. The guiding principles have to be clearly laid down to ensure that such institutional mechanism is autonomous and steers away from a top down style.

Establishing Economic Value

The suggested move towards full cost recovery of water used as the means for achieving efficient use of water and phasing out subsidies in the longer term is logical. This, if implemented once the basic drinking and hygiene needs of human beings are met is the fundamental restructuring that this sector requires at this stage. The policy does reasonably well in echoing this once again but unfortunately does not comment about the timelines predicted to set up institutional procedures in the states. As discussed earlier, developing a model that works towards establishing economic value of the resource and also wants to ensure the fundamental rights is fraught with conflicts and challenges. These theories are absolutely at the opposite ends of the economic spectrum and in the context of a growing country like India, have their pros and cons. While the fundamental rights concept has an advantage in the form of universality, it has failed in practical application. The strong point of the full-cost concept[4] is its upfront call to action to recover the full costs from consumers, and a strong belief that water and sanitation problems will then be taken care of by the market forces. The critics of the concept however are not convinced about the market forces' ability to allocate sufficient supply of resources to users with a very limited capacity to shell out the costs. One of the programmes that may help bridge this gap between full cost recovery and supply to the economically vulnerable segments is small loans from microfinance institutions. Given a suitable programme model, this financial assistance can act as an alternative to the initiatives by the public sector or large partnership projects. As the subsidies through tax transfers and grants become more and more difficult to obtain, this form of financing can help ensure extending the services. India has a tremendous opportunity to combine appropriate cost-recovery mechanisms from different segments and support the vulnerable segments with innovative solutions.

More Carrots, Less Stick

Interestingly, with regard to water quality conservation, India's national water policy departs from the 'polluter pays principle', that has been the traditional approach. It now intends to incentivize better management of effluent treatment and recycling of water. The water thus reclaimed can prove to be a significant resource to curb the growing deficit. The 'carrot instead of the stick' approach may also foster an environment of technological innovation. This contrasts with the strategies adopted by neighbouring China, where the government's policy reactions to water quality concerns mostly count on reinforcing the supervising capabilities and enforcement mechanisms. Increased penalties including stiff fines for the polluting entities, as stipulated by the 2008 Water Pollution Law has strengthened the existing legislation. In order to expand the supervising capabilities Regional Supervision Centres have been established and Water Quality Bureaus have been set up throughout the country to keep an eye on the pollution levels. The Chinese 'three red lines' policy has been expanded to cover a wide range of organic and microbial pollutants as well as concentrations of heavy metals. Further, new requirements have been introduced that stipulate that 95 percent of tested water must meet national water quality guidelines. The administrative augmentation has been well rounded by continuing regulatory reforms at the same pace. In other Asian countries like Malaysia where there have been trends in rising water pollution, the administrative authorities have responded with a stronger monitoring network and analytical capabilities. India may take a more positive approach and should definitely endorse the model that promotes innovation towards water treatment and waste management, but it should continue to empower a structure that keeps an eye on large polluters and their impact on the communities and the ecosystem.

China's Grand Strategy

India's neighbour faces similar development challenges—dense population and limitations of available water resources. Things get even more complex in China, as it encompasses a blend of geographical, political, economic, and social dimensions. In a nutshell, China's water resource challenge consists of both water quantity and quality issues, each of which has presented unique challenges for the Chinese policy[5]. Although it has the risk of being destabilized by the intergovernmental conflicts and corruption, the Chinese government is implementing perhaps the world's boldest water resource management strategy that will support economic development. It ultimately aims to transfer some 45 billion cubic meters of water per year from central and southwest China to augment the flow of the Yellow River and meet urban water demand in the Beijing-Tianjin region. Although its completion is expected to meet projected water demand growth in the north, thus rectifying China's fundamental geographic inequality in water availability, its outlay is colossal. Although the approximations suggest a number close to $ 62 billion, the final cost is likely to be prominently higher. The cost of this magnitude cannot be absorbed by the state and is proposed to be recovered in full through major transformations in tariff structure. Another great example can be seen in Singapore's water management that distinctly shows that basic utilities do not have to be subsidized or discounted for better access. The country has improved both long term security and good quality of water by integrating ecological costs, effectively tackling water scarcity and making its water sector more market driven. It has also introduced separate tax rebates on utilities and subsidies directed at lower income households in order to separate distributive impacts from over-utilization.

Walking the Talk

Coming back to India, a few inadequacies notwithstanding, the draft National Water Policy, 2012 at least works as a step ahead in the direction of better management of water resources in several ways. The draft introduces various progressive features but could have been less ambiguous in some. There was an opportunity to distinctly lay down some basic principles on the allocation of water resources. At the central level it will be important to set some expectations and national perspectives, even as the states can thereafter prioritize allocation in line with its agro-climatic, domestic or industry needs. The nation's water policy needs to ascertain that there is a greater degree of consistency in water management across the country and equity for different users and within same user group. If the policy stands by its statement on water as a fundamental right, it should ensure that the state takes up the resultant role to provide this to the citizens.

The country would need a fully integrated approach to enhance water available for use, manage demand and resource usage efficiency. All measures that would manage demand or enhance water usage efficiency would automatically lead to enhanced water availability for usage. Millions of people in Asia and the Pacific could gain access to clean water if leaks were plugged and water utility reforms adopted, a study by the Manila-based Asian Development Bank (ADB)[6] has stated. ADB estimates that 29 billion cubic meters of water is lost each year in the region. At the water planning stage if we know the gap between demand and supply, different strategies of water conservation, water augmentation, water recycle and re-use can be employed depending upon the specific conditions and requirement of a particular region to bridge the gap.

Another requisite is that water policy should establish strong inter-connects with other state and central policies, i.e., industrial policy, agricultural policy, land use policy and mining policy. These need to be emphasized and discussed widely to enable congruent stipulations and unambiguous interpretation.

New technologies deployed towards better management of water resource will play a large part in solving the water scarcity. In order to make them mainstream, corporate and investors will not only have to work more closely with each other, but also engage with other stakeholders to remove negative perceptions and biases. We can take a lesson from Singapore for its remarkable success in tackling its water crisis through effectual water policy reforms[7]. As it was faced with severe water shortage the island country has been working on continuous reforms through the last three decades. Singapore's model not only reveals the significance of political will in revolutionizing water policy, it also exemplifies the important roles of institutional integration, efficient land use planning, enforcement of policies, public awareness initiatives and the application of water-saving technology. Singapore has also had an outstanding success in reducing demand for water resources since implementing the block-rate pricing policy. There are other nations like South Africa where the success is limited but the policy reforms have played the central role in addressing the perilous water crisis[8]. The National Water Act of South Africa has made noticeable advancement in narrowing the gap of available water since its implementation. It is true that the actual quality of the drinking water is still not up to the mark and the country has a long way to go with regard to water policy but its emphasis on water equity has had deep impacts on the society of the country. The number of people who now have access to water has increased by more than ten million since 1994.

Industry associations in India and investors from all over the world have been affirmative towards the new policy. There are some apprehensions from industry segments on the resource allocations and states such as Bihar, Punjab, Kerala and Madhya Pradesh have expressed reservations on some facets of the policy as well. The policymakers and the ministry need to assure these segments about the wider interest, fairness and mechanisms for better resolutions.

Water or the lack of it could become a major limiting factor to social and economic growth of the country. Progress has been made in the general direction of national legal framework of principles on water, which, in turn, would pave the way for essential legislation on water governance in the states. There is a compelling need to rise above political, ideological and regional differences and also to steer clear of the short term project centric approach. All the players should adopt a longer term, all-inclusive approach that helps in executing the policy on managing issues around water resources. As the great Thomas Edison once said—'Vision without execution is hallucination.'

Pricing the Elixir

'When the well's dry, we know the worth of water.'

—Benjamin Franklin

It may not strike tomorrow but the peril is real nonetheless. The looming crisis of our diminishing water resources poses a catastrophic situation as grave as any that we have ever faced. Perhaps, no other country can understand the water crunch better than India. The rapid growth in population and urbanization; the resultant demand for industrial and agricultural goods, coupled with inefficiency and pollution have put a huge pressure on the already dwindling supply of fresh water. The estimates bring to light the fact that by 2020, the country's demand for fresh water will exceed all sources of supply. This situation left to its own will lead to a severe drop in per capita availability of water—well below the critical 1,000 CM—by 2025.

Fundamental theories of economics will tell us that a resource that finite, already scarce and without any substitute should be a highly priced commodity. Add to that the large scale investment that needs to be made to build water infrastructure over the coming years we have a compelling case for relooking at the commodity's valuation and pricing. On the contrary, the collective approach towards water resources in general and towards water pricing in particular has been somewhat unsustainable.

What complicates the simple logic, especially (but not exclusively) in the Indian context is that water is not considered purely as an economic good. It is also regarded as a social good as access to clean water is vital to people's wellbeing; and it also has cultural and religious values embedded deeply into the rituals and imageries. The lasting debate between indisputable rights of humans and other living species and appropriate valuation in a free market has been one of the major reasons why the actual or full cost pricing has not been implemented in this sector. The result of this perplexity is that across the user types we continue to pay some of the lowest water prices on the globe. Not only has this given us an erroneous sense of water's real value, it also has not provided for much required maintenance of and further investment in water infrastructure.

All over the globe, water pricing is being implemented with the dual aim of expanding supply and encouraging more responsible use[1,2]. Major economies are pushing for significant increases in the price of water around the world as concern mounts about dwindling supplies and rising population. As the costs within the water sector are rising faster than the prevalent rate of inflation, full cost pricing is a progressively accepted pricing structure for drinking water and wastewater service which fully recovers the cost of providing that service in an economically efficient, environmentally sound, and socially acceptable manner, and which promotes efficient water use by customers. Full costs may include source protection costs, operating costs, financing costs, renewal and replacement costs and improvement costs associated with extracting, treating or distributing water to the public and such other costs which may be specified by regulation. Full cost recovery from tariffs which may theoretically be the ideal solution, in practice remains a distant objective in many countries. However, even very poor countries can reach important cost-recovery targets at the sub-sector levels.

In many countries, commoditization is converting water from a non-market good to a market good. Several countries have also experimented with transferring the management and entire control of water organizations to private sector in an effort to improve the overall system. Although there is the belief that private sector can mobilize capital faster and can bring in efficiencies, there have been concerns about social inequities, ecosystem impacts and water quality.

Draft National Water Policy: A Nudge to Pricing Reforms?

Ministry of Water Resources, Government of India, *released the National Water Policy in 2012*[3] *and invited public feedback following which it has been put up as a revised draft. Even as the policy itself attempts to cover varied subjects including the legal frameworks, demand management, usage efficiencies, infrastructure and skill development, it also touches upon the issue of water pricing in a brief way. The key action items enlisted under this head are—*

- For the pre-emptive and high priority uses of water for sustaining life and ecosystem for ensuring food security and supporting livelihood for the poor, the principle of differential pricing may have to be retained. Over and above these uses, water should increasingly be subjected to allocation and pricing on economic principles
- A Water Regulatory Authority (WRA) should be established in each state. The Authority, inter alia, will fix and regulate the water tariff system and charges, in general, according to the principles stated in this policy in an autonomous manner. Such tariff will be periodically reviewed
- In order to meet equity, efficiency and economic principles, the water charges should preferably/as a rule be determined on volumetric basis

- Recycle and reuse of water, after treatment to specified standards, should also be incentivized through a properly planned tariff system
- Water Users Associations (WUAs) should be given statutory powers to collect and retain a portion of water charges, manage the volumetric quantum of water allotted to them and maintain the distribution system in their jurisdiction. WUAs should be given the freedom to fix rates subject to floor rates determined by WRAs
- Heavy under-pricing of electricity leads to wasteful use of both electricity and water. This needs to be reversed. As an alternative, where limited ground water use for agriculture at a subsidized cost is considered desirable, separate electric feeders for such a use should be considered

Although the intent of pricing water right seems to be in place, what needs to be prioritized is how the proposed system translates the same into practice. As India is not the only nation caught in this debate and state of confusion, it can take cues from some nations who have or are now trying to look at the issue right in the eyes. Many of these policy and change makers are aware that increases in prices will be—and in some countries are already proving to be—hugely controversial, they feel that they can be managed without negatively impacting the poorest. They also concur that as long as most countries provide huge subsidies for water it will not be possible to change the wasteful habits of consumers, farmers and industry, nor to raise the investment needed to repair old supply systems and build new ones.

The existing water tariff rates as per Delhi Government are as follows:

Service Charges		
Category	**Nature of Premises**	**Charges per month (Rs)**
C-I (Domestic)	Premises with built up area upto 200 sq.m. Premises with built up area above 200 sq.m	Rs. 40 Rs. 120
C-II (Non-domestic)	Commercial	Rs. 250
C-III (Non-domestic)	Industrial	Rs. 600

Source: Citizen Charter, Delhi Jal Board, Website of Government of NCT of Delhi; details available at: http://delhijalboard.nic.in/djbdocs/about_us/charter.htm

Volumetric Water Charges (based on consumption)	
Consumer category/ Consumption Slabs	**Volumetric charges (Rs/Kl)**
Category I - Domestic	
Upto 6 kls	0.00
7-20 kls	2.00
21-30 kls	7.00
31 kls & above	10.00
Category II – Commercial	
Upto 25 kls	10.00
26-50 kls	20.00
51 kls & above	30.00
Category III – Industrial	
Upto 25 kls	15.00
26-50 kls	25.00

51-100 kls	35.00
101 kls & above	50.00

Source: Citizen Charter, Delhi Jal Board, Website of Government of NCT of Delhi; details available at: http://delhijalboard.nic.in/djbdocs/about_us/charter.htm

Note: 1.50% of volumetric water consumption charges are recoverable in every category

When China Gets Serious

Neighbouring China, with many of its cities short of water, has been encouraging wastage in the recent past by selling it at heavily subsidized prices. Latest moves by several cities to bring in higher tariffs have shown how sensitive water consumers can be about rising prices, even at what seemed an opportune time. Not daunted by the adverse reaction—even from the factions within state owned media—the country's water conservancy plan co-distributed by National Development and Reform Commission (NDRC), continues on its path to include a system of progressive pricing[4, 5], to discourage excessive consumption. While on one hand, China confronts the intensifying water crisis through massive infrastructure development aimed at securing supply, it has also taken up the delicate task of moderating demand. There still remains the complication of avoiding excessive burden on the farmers but pricing is still being looked at as an effective strategy. For rural residents, the country will explore ways that give them price discounts when their water use is within set quotas and calculate prices progressively when they use more than the quotas. The pricing reforms are part of the government efforts to conserve resources and make their prices reflect market demand amidst growing supply pressures.

The key ministries and think tanks remain firmly on the side to raise tariffs. Earlier this year, NDRC, the Ministry of Finance and Ministry of Water Resources issued a joint standard on Water Resources Fee (WRF) which includes—

- a clear water resources tariff formulation system
- different standards for different type of water resources such as surface water and ground water
- reasonable water tariff standards to promote water resources tariff reform, considering local water resources condition, local economic development and users in different industries
- strict control on underground water exploitation
- support on agricultural water consumption by exempting or lowering water tariff within a certain consumption range
- encouragement of water recycling
- reasonable water tariff standards for hydropower industry
- punitive standards for excessive water consumption
- Management and control on excessive water tariff collection

Clearly, the Chinese government has understood the urgency of the water issues and is setting the prices to encourage more efficient use and protection of the nation's already scarce water resources.

South Africa Readies for Water Challenge

According to the Ministry of Water and Environmental Affairs South Africa will need R570-billion for investment in the water value chain in the next ten years[6]. The money is required to finance water resources infrastructure, conservation and demand management nationally and across the existing water boards. The vast funding needs to raise the probability of upward revision in water pricing for various types of consumers in the coming years. The country's water resource pricing strategy is based on the following principles—

- **Social Equity**, which is focussed on redressing the imbalances of the past with respect to the inequitable access to basic water services at affordable tariffs within municipal areas, by facilitating a conditional subsidy on raw water cost where stepped tariffs are introduced; and inequitable access to water for productive use purposes by subsidising tariffs for emerging farmers for a limited time period
- **Ecological Sustainability**, which requires safeguarding the ecological reserve; water quality protection; and water conservation and demand management
- **Financial Sustainability** aimed at generating adequate revenue for funding the costs related to the management of water resources and the operation, maintenance and refurbishment of existing schemes
- **Economic Efficiency**, which aims at promoting the efficient allocation and beneficial use of water, i.e., water should be priced at its opportunity cost; and at providing for administrative as well as market-related measures to achieve this goal

Overall, the strategy is aimed at moving towards tariffs which recover the full economic costs of providing raw water from the resource. However, poorer domestic consumers and emerging farmers will continue to be subsidized and the cost-reflective tariffs will be phased over time.

One lesson that other countries including India can learn from the South Africa experience is that it is important to recognize that not all water use should be regarded as equal. The step-pricing approach, factors in the fundamental human right of access to potable water and at the same time looks at like swimming pools, gardening etc. in a different way. Responsible usage is not the only benefit arising due to this full cost recovery as it also helps provide crucial funds to repair the stressed infrastructure.

Not Surprising—Singapore Gets it Right

The proactive measures taken by Singapore[7] are characterized by a number of achievements over the decades in ensuring access to affordable and high quality water in challenging conditions is commendable. Apart from the innovative approach and infrastructural support to integrated water management including reuse of reclaimed water, the establishment of protected areas in urban rainwater catchments and seawater desalination, the country has put strong emphasis on water pricing and educating the citizens. The significant investment in the sector is supported by pioneering financing mechanisms and tariff structure.

Water Tariff in Singapore (Source PUB, Singapore's National Water Agency Website)			
Tariff Category	**Consumption Block (m3 per month)**	**Tariff ($/m2) [before GST]**	**Water Conservation Tax (% of tariff) [before GST]**
Domestic	0 to 40	1.1700	30
	Above 400	1.4000	45
Non-Domestic	All units	1.1700	30
Shipping	All units	1.9200	30

The scarcity value of the precious resource reflects in the water pricing in Singapore. The amount of water consumed forms the basis for the water tariff. The Water Conservation Tax (a percentage of total water consumed) introduced in 1991, for instance, reinforces the message that every drop of water is precious and everyone must do their part to conserve water.

At the same time, the authorities have recognized that help may be needed at low-income levels. Under the U-Save programme there are rebates to help offset utility bills, including water expenses. From 2007 to 2011, a household staying in one to three room flats received an annual U-Save rebate of $110 to $360 (or average about $10 to $30 per month), compared to the average water bill of less than $35 a month. This dual approach of resource conservation and directed assistance ensures longer term access to affordable, high quality water by all the citizens.

Turning It Around the Australian Way

The Council of Australian Government (COAG) run water reforms in Australia commenced from 1994[8]. Prior to that, while there were evidences of some sort of reform at jurisdiction levels, the administrative expediency often overcame commercial sense in pricing of water supply and services. This approach led to a number of problems, including—

- use of water without regard to its cost of supply, leading to excessive consumption, and to environmental impacts and the need for costly investments in new supply capacity, if left unchecked
- under-recovery in the costs of service provision and major asset refurbishment needs (particularly in rural areas) for which adequate financial provision had not been made
- service-delivery inefficiencies and a lack of incentive to provide reasonable levels of service at lowest possible cost

- commercial and industrial water users often paying far more than the cost of service provision (and cross-subsidizing domestic water customers) because of property rate-based charges

The accumulation of these challenges not only turned into starting points for COAG's water pricing agenda but also echoed the broader micro-economic issues and the modifications trained at improving social wellbeing and economic productivity.

Over two decades, these pricing reforms have delivered on several parameters. The usage-based tariff in most urban areas has steered efficient consumption of water and consistent reduction in overall residential water consumption. Independent economic regulation of urban and rural water services has not only provided better analysis of water businesses' expenditure but has brought in significant unambiguousness and ownership for all the participants. Secondly, the metropolitan and urban water businesses are much better placed to commit funds for new investments in water infrastructure from the customer revenue due to this drive to recover full efficient costs. Further, shifting fixed network access fees from entitlement-based charges to charges based on delivery rights, combined with the removal of exit fees, has helped the rural water market drive efficiency in water use.

In Proactiveness Lies the Answer

In many parts of the world the precious resource is not really scarce at present, but there is already a growing recognition of the immense value of water fuelled by concerns over potentially shrinking supply, growing demand and threats to water quality. Compared to other countries, for instance, Canada is blessed with an abundance of fresh water[9]. The country along with Iceland, Norway and New Zealand stand out as having the largest supply of renewable freshwater per capita in the world. That is not stopping the policy community from training their focus on market-based mechanism—specifically pricing—for effective resource management and its apportionment[10]. Imparting price to water may serve as an effective mechanism to encourage conservation, efficient allocation amongst diverse user segments and reduce wastage. France adopted the principle 'water pays for water' very early[11] and has been practicing it for several decades now. To implement this principle, the six water agencies make use of a wide number of mechanisms. The country has founded a vigorous financing framework that includes well-defined codes, a wide range pricing vehicles, and robust institutions to execute the policies and best practices. Further, the country continues to experiment and facilitate improvement in the system to make it match the evolving ecological sustainability dimension. The national water laws in the country also applies 'polluter user pays' under which the River Basin Agencies can directly determine charges at basin level for withdrawals and discharges.

One of the interesting trends to observe on global water pricing is the development in the US[12].

New York City Water & Sewer Historical Rates; Source NYC Water Board Website

Fiscal Year	Period Covered	Charged in metered water	Water Rate per 100 of (748 Gsls)	Sewer Rate per 100 of (748 Gals)		Total w/s
2014	7/1/13-6/30/14	+ 5.6%	$3.58	159% of water =	$5.69	=$9.27
2013	7/1/12-6/30/13	+ 7.0%	$3.39	159% of water =	$5.39	=$8.78
2012	7/1/11-6/30/12	+ 7.5%	$3.17	159% of water =	$5.04	=$8.21

2011	7/1/10-6/30/11	+ 12.9%	$2.95	159% of water =	$4.69	=$7.64
2010	7/1/09-6/30/10	+ 12.9%	$2.61	159% of water =	$4.15	=$6.76
2009	7/1/08-6/30/09	+ 14.5%	$2.31	159% of water =	$3.67	=$5.98
2008	7/1/07-6/30/08	+ 11.5%	$2.02	159% of water =	$3.21	=$5.23
2007	7/1/06-6/30/07	+ 9.4%	$1.81	159% of water =	$2.88	=$4.59
2006	7/1/05-6/30/06	+ 3%	$1.65	159% of water =	$2.62	=$4.27
2005	7/1/04-6/30/05	+ 5.5%	$1.60	159% of water =	$2.54	=$4.14
2004	7/1/03-6/30/04	+ 5.5%	$1.52	159% of water =	$2.42	=$3.94
2003	7/1/02-6/30/03	+ 6.5%	$1.44	159% of water =	$2.29	=$3.73

With not much stress on the access to fresh water, the consumers have traditionally used twice the amount of water that European consumers use, while paying much less. With time, however, there has been significant investment to develop new water resources through desalination, water reuse and the acquisition of water rights. The variable aspect of the cost of water has consistently been rising. Some of the important cities in the west coast, such as San Diego, Los Angeles and Las Vegas, are already adopting the European approach to water pricing. Even then, it is quite disconcerting to note that the negative correlation between water availability and water price still exists. In hot and arid regions where the usage is much higher the users tend to pay little for water—almost free in Libya, Turkmenistan and heavily subsidized in Saudi Arabia, India and Egypt. Pricing aggravates this inappropriate pattern of demand.

In most developed countries, there is now a well-defined trend towards volumetric charging or higher charge for higher usage[13]. Even as the fixed charges still exist in some countries, large free allowances are on a decline. Further, the trend also indicates a growing adoption of volumetric charging towards increasing block and away from the decreasing block thereby incentivizing lower consumption of water.

Globally, the pricing mechanism for water supply is somewhat less complicated than it is for wastewater treatment. In addition to the confusion with regard to the responsibilities amongst various entities, the use of water directly from natural sources in the environment creates further complications. As the effluents' content varies greatly from industry to industry, it is broadly understood that industrial water consumption levels may not accurately represent the sewerage and sewage disposal costs. Many countries therefore are opting for separating industrial water usage charges from wastewater charges.

In case a particular industrial effluent requires extra capacity to process the same, there is usually an additional charge over and above the standard sewerage charges in most of the countries. Pollution content may also determine the industrial effluent charges. In France, for example, a charge is levied on the eight types of pollutants—including heavy metals, phosphorus, soluble salts, etc.—deemed most dangerous and difficult to treat . The charge is calculated as a function of pollution produced during the period of maximum activity on a normal day.

One of the major reasons for the rising charges in most countries in recent years is that quality of water (particularly ground water) is worsening as a result of over-usage and other inappropriate consumption habits. The treatment of water of this poor quality requires more sophisticated and costly treatment. As the government budgets get stretched, increase in water prices is only logical. Therefore, more than across-the-board subsidies there is a growing stress on affordability and reasonable approaches for achieving social goals. Range of policy measures have been developed by various countries that are aimed at resolving the affordability problems of low income groups while meeting the overall environmental and economic objective.

Even in the international context, the pricing structures reflecting the full costs of providing the service, was easier to achieve in case of municipal and industrial water. In many countries, water used for agricultural purposes continues to be broadly subsidized, which discourages efficient use of the already scarce resource. There may be some movement towards full recovery of infrastructure and operating costs from users in this group, but the same is extremely slow and needs greater public support for speedier reforms.

Execution Will Win the Game

A well conceptualized pricing strategy and its effectual implementation can set the nation's water policy and the sector reforms on a virtuous course—satisfied consumers and rewarded investors. India needs to get its priorities in water pricing right. The improvements on the demand side as well as all the conservation endeavours will get a significant boost once the enabling structure of pricing is firmly in position. Administering the pricing strategy across the user categories and ensuring that the amounts due are collected requires further political will.

Enough is already said about the 'economic good' that it is. In reality, the water pricing strategy will also need to address all the other 'goods' that it is believed to be. It has to win the trust of the society as well as the government. As in most circumstances the end user of water services has no choice of supplier, active engagement with the customers may help dilute the apprehensions. The overall pricing mechanism should be able to clearly demonstrate the benefits to the larger society and be competent in handling extra-sensitive issues like services to the vulnerable groups.

Secondly, even as innovative pricing mechanisms can help the sector raise much needed funds for infrastructure development, it has to be complemented by sound technical understanding of contemporary technologies and an astute ability to analyse how the alternative investments add value to efficiency and quality enhancement. Successful strategic plans for the water sector should clearly emphasize on the prospects of reducing costs. Adoption of lower cost technologies, deadlines for attaining targets and rationalizing construction and environmental standards are some of the important opportunities that are linked to policy decisions. Although largely dependent also on local conditions and administrative standards, these could play an important role in improving the operational efficiency of utilities.

Although a lot needs to be done, what may work for India is a phased but firm approach in renovating and building water infrastructure, assembling strong evidences of clear benefits in terms of better access and quality. Balancing financial, environmental and social objectives in water pricing policies remains a tough act but in order to make sustainable progress, the pricing mechanism has to reflect fairness, transparency and a demonstrated interest in long term innovation.

The water pricing needs to consider the appropriate mix of pricing instruments from a wide range available today. Taking a cue from some of the countries such as France and South Africa, it can implement multiple water-related taxes and water-related charges. Effective financial planning for the water sector requires finding the right mix of revenues from what are popularly referred to as the 3Ts—tariffs, taxes and transfers[14], with the right institutional framework in place these levies can be further segregated to clarify what is being asked for in return of a service and what is being taxed. They may also distinguish between various types including regulatory levies, water use levies, water pollution levies, water service levies, and fines and damage compensation penalties. The strategy, at the same time, can also develop the relatively novel categories of negotiated pricing instruments, with the instruments such as water abstraction rights, water pollution rights, or wetland development rights—water authorities do not receive revenues to fund the resource management actions, while the private actors that receive the revenues may allocate appropriately to fund such actions.

Water in India is practically free or available at throw-away price. The extreme argument that water is fundamental to life and hence cannot be bought and sold does not hold much water as food products are marketable and priced. There is a cost associated with water supply and with maintaining infrastructure funding, which will not be sustainable for government budgets. Somewhere in these definitions of 'goods' and blending in the aforesaid priorities, the country needs to explore the appropriate price at which the fundamental right to water becomes a workable proposition. And these proposals need to be replicated to agriculture as well, where a great difference can be made. With agricultural production projected to expand in the coming decades to feed the growing population, farmers will need to improve water efficiency. The farmers should be encouraged to pay not only the operation and maintenance costs for water but also a fair share of the capital costs of water infrastructure. A gradual rise in water prices would also improve efficiency in this area and even a marginal saving in irrigation water usage can release substantial amounts of water for agricultural expansion as well as for meeting the needs of other sectors like domestic water demand. A recent OECD report on water management in agriculture, for example, has shown that, in areas where the price of water for agriculture has increased, agricultural production has been unaffected. Another area for stringent action is pollution control norms which would encourage industries to treat and reuse water.

Water pricing is for real—in India it has to be initiated through active policy intervention that may empower the natural market forces in due course. Water is a resource that has to be managed and a source of services that come at a cost. At the core of water pricing lies the clear understanding of how these costs are shared among different categories of users and beneficiaries and are fully covered. In developing and transition economies like India where the deterioration is already rampant that may lead to the eventual collapse of infrastructure, water tariffs should become over time the main source of finance. If handled correctly, in a customer-oriented way, in the long term the customers will generally be willing to pay the cost of good quality and efficient services that are affordable and respond to their needs. The emerging consent is that water policy needs to pinpoint what objectives are financially realistic, taking into account that there are only finite sources of revenue. Sustainable management of India's water resources need significant financial resources and water pricing has that potential to raise the same consistently. As funds from public budgets and grants in our country become stretched and less predictable, water pricing may be the main source of revenue for India's water sector—just as it is for countries like France and the Netherlands.

While there is no doubt that the water challenge is daunting, it is also true that being an alarmist alone would not help. Taking some firm steps towards adoption of water pricing by volume appropriately at the municipal, industrial and also agricultural level will help aid water conservation, improve supply and pay for crucial infrastructure.

Virtual Web of Water Trade

'To master the virtual equation and make all the elements work together, you have to become the connector. In fact, your greatest role...is to link the various parts...'

—Yael Zofi

The concept of water trade has basically evolved as water is becoming a scarce commodity with time. Water woes have assumed enormous proportions.

International trade of commodities carries flows of virtual water over large distances. Still emerging 'Virtual Water' concept, defined as volume of water required to produce a commodity or service, is directly linked to the water footprint concept. Water footprint, however, is a much broader concept and considers the location, time and source of water used.

The development of water trade, based on expanding water footprint, in the economy has emerged since there exists an urgent need for an indicator of water use in relation to water consumption. This, in turn, is expected to facilitate the process of assigning value to the intangible and relatively ambiguous process of water usage and to capture water exploitation.

The concepts of water footprints and virtual water are useful in describing the relationship among water management, international trade and politics, national and international policies and use of water resources as it pertains to human consumption. These are useful in illustrating the true influence of economic activities on water. Greater awareness should be accompanied by measures to improve water productivity ('output per drop') in water-stressed environments and to reduce the polluting side effects of production[1].

There is a subtle difference between the concepts of water footprint and virtual water, as both these concepts imply measurement of the volume of water. However, water footprint has a wider application. Unlike virtual water, which accounts for only the volume of water in a good or a service, water footprint can also be referred to a consumer (by measuring the water footprint of goods and services used by the consumer) or to a producer (by measuring the water footprint of goods and services produced by the producer). Furthermore, water footprint also states the spatial and temporal details of water usage and also the source of water used.

Virtual Water

Virtual water, closely linked to the water footprint concept, is defined as the volume of water required to produce a commodity or service. The concept was introduced by Allan John Anthony, a British geographer, awarded the Stockholm Water Prize 2008, in the early 1990s, when studying the option of importing virtual water (as opposed to real water), as a partial solution to problems of water scarcity in the Middle East. Virtual water trade has been promoted as a tool to address national and regional water scarcity issues since that time. In the context of international (food) trade, this concept has been applied with a view to optimize the flow of commodities considering the water endowments of nations. Allan elaborated on the idea of using virtual water import (coming along with food imports) as a tool to release the pressure on the scarcely available domestic water resources[2].

Virtual water import thus becomes an alternative water source, next to indigenous water sources. Imported virtual water has therefore also been called 'exogenous water'. When assessing the water footprint of a nation, it is essential to quantify the flows of virtual water leaving and entering the country. If one takes the use of domestic water resources as a starting point for the assessment of a nation's water footprint, one should subtract the virtual water flows that leave the country and add the virtual water flows that enter the country[3].

Virtual Water is a tool for determining the movement of water through international trade[4]. This process, though criticized for being crude and too simplistic, helps environmentalists to place a value on the ambiguity that surrounds the estimation of fast-depleting water resources, within a nation. International water dependency is as significant as domestic water dependency. One of the major revelations supporting this statement is that approximately 16 percent of the global water used is not for producing domestically consumed products but products for export. With increasing globalization of trade, global water interdependencies are most likely to get augmented, thus underlining the need for inclusion of such estimates at the international level especially by global trade bodies and multinational organizations.

As regards domestic water, it is meaningful to realize that the existing pattern of inter-state virtual water trade in India is exacerbating scarcities in already water scarce states, with virtual water flows moving from water scarce to water rich regions and running in the opposite direction to the proposed physical transfers of goods. Because water is heavy relative to its value, it is not feasible to transport it in bulk over long distances, with the exception of limited schemes for drinking water. Thus, water is predominantly a local concern, although it becomes a regional issue where rivers or lakes cross national boundaries.

International trade in agricultural commodities depends on a lot more factors than differences in water scarcity in the trading nations, such as differences in availability of land, labour, knowledge and capital and differences in economic productivities in various sectors[5]. Also the existence of domestic subsidies, export subsidies or import taxes in the trading nations may influence the trade pattern. As a consequence, international virtual water transfers cannot at all or only partially be explained on the basis of relative water abundances or shortages. Virtual water is not only relevant for considering the demand and supply of water, but also with respect to commodities trade. It is not about economies of scale in terms of production and trade, but embraces a broader perspective. An inverse relationship can be established between a country's cereal import and its per capita renewable water resources[6].

The total volume of water used in the agricultural sector is calculated on the total volume of crop produced and its corresponding virtual water content. Further, the virtual water content (m^3/ton) of live animals is calculated based on the virtual water content of their feed and the volumes of drinking and service water consumed during their lifetime. The calculation of the virtual water content of livestock products is again based on product fractions and value fractions. It is seen that the higher we go up in the product chain, the higher will be the virtual water content of the product. For example, the global average virtual water content of maize, wheat and rice (husked) is estimated at 900, 1300 and 3000m3/ton respectively, whereas the virtual water content of chicken, meat, pork and beef is estimated at 3900, 4900 and 15500m3/ton respectively.

However, the virtual water content of products strongly varies from place to place, depending upon the climate, technology adopted for farming and corresponding yields[7]. In terms of trade, virtual water flows between nations have been calculated by multiplying commodity trade flows by their associated virtual water content[8]. The patterns of inter-state virtual water trade in India and global food trade trends discussed by De Fraiture[9] et, al show that water endowments alone are unable to explain the direction and magnitude of trade. The implicit assumption behind measuring every commodity by its virtual water content is that water is the most critical and scarcest resource input. However, this assumption does not always hold. There are several key inputs that go into the production of food and these other 'factors of production' might tilt the balance of decisions against the logic of virtual water which dictates water saving as the sole criterion[10].

The virtual water content of an industrial product is calculated in a similar way; as described earlier for agricultural products. There are however numerous categories of industrial products with a diverse range of production methods, and detailed standardized national statistics related to the production and consumption of industrial products which are hard to find. As the global volume of water used in the industrial sector is only $716Gm^3/yr$ ($\approx$10 percent of total global water use), there is—per country estimate—simply calculated an average virtual water content per dollar added value in the industrial sector ($m^3/\$$) as the ratio of the industrial water withdrawal (m^3/yr) in a country to the total added value of the industrial sector ($/yr), which forms a component of the Gross Domestic Product[11].

A number of alternatives have been suggested including decentralized water harvesting and artificial recharge of aquifers, improving the productivity of agriculture in water scarce regions (which, it is claimed, continues to waste precious water resources), improving the efficiency of India's public irrigation systems through involvement of stakeholders in the management of irrigation, and using virtual water trade is one of them[12]. The stakeholders here are referred to as companies taking initiative for sustainable development. In the Indian context, many companies are beginning to understand the need to measure their water footprint, including that of their supply chains, and to relieve water stress in the communities where they operate.

Future Movements of Virtual Water

Like the global scenario, the existing pattern of inter-state virtual water trade in India is exacerbating scarcities in already water scarce states, with virtual water flows moving from water scarce to water rich regions, thereby running in the opposite direction to the proposed physical transfers[13]. Rather than being dictated by water endowments, virtual water flows are influenced by many other factors such as per capita availability of arable land and more importantly by biases in food and agriculture policies of the Government of India as indicated by the Food Corporation of India's procurement patterns. In order to have a comprehensive understanding of virtual water trade, non-water factors of production also need to be taken into consideration[14].

Virtual water trade in the year 2000 accounted for one fourth of the global virtual water budget, precisely 26 percent[15]. This importance is likely to dramatically increase as projections show that food trade will increase rapidly—doubling for cereals and tripling for meat between 1993 and 2020[16].

The relatively large volume of international virtual water flows and the associated external water dependencies strengthen the argument for putting the issue of water scarcity in a global context[17].

Major food exporting countries, overall, have low irrigation intensity[18]. The proportion of food production from irrigated areas is considerably small. The global virtual water trade is dominated by green water. Such a trade is efficient in terms of the opportunity cost of water used. However, the high water productivity in the major exporting countries is partly due to the high inputs of chemical fertilizers and pesticides[19]. Given the increasing pressure on the global blue water resources, more effectively utilizing green water may have to be a direction which the world agriculture ought to pursue in the near future[20].

Industries are increasingly traversing the way towards minimizing their water footprint, as they have realized that with this they will only derive benefits from and reduced dependence on state supply, in the process reducing their cost of use of water.

Virtual water is an intensive concept in so far as it links water, both freshwater and soil water, in the productive process of crop production. Here water and food have been linked in the activity of crop production. Implicit is the idea that crop production, and therefore indigenous food security, can be limited locally by the availability of water[21].

Further, the concept of the shadow price is not diminished by its limitation to a theoretical role. The concept of the shadow price remains very valuable in providing an analytical perspective on underlying fundamentals. Similarly, the concept of virtual water, it could be argued provides an analytical perspective on how economies ought to achieve water security[22].

Bottle with a Narrow Neck

'We shape our buildings; thereafter they shape us.'

—Winston Churchill

The state of infrastructure is the principal yardstick for judging a nation or a region's development. These complex physical structures and services provide the much needed support for enterprise, society and economy to function effectively. Several sets of interrelated structural elements are combined to build frameworks that reinforce the overall edifice of development.

Infrastructure facilitates the production of goods and services—distributing finished products to the markets through efficient transport, supplying power to industrial and residential consumers, connecting the society through widespread telecommunication network and so forth. Basic services, considered as social, such as school and hospitals are also covered under this sector. In spite of the claims of a 'shining economy', India continues to grapple with huge infrastructure related problems. There is in fact hardly any need for detailed analysis of data from various sources—a week of living in the country or travelling in any direction for a couple of hours will make these challenges quite obvious.

Water supply and sanitation are important elements of infrastructure as they involve not only the intricate mix of natural resources and physical infrastructure, but also as the verdict is still out on whether they are essential commodities or social service. Infrastructure systems include both the fixed assets, and the control systems and software required to operate, manage and monitor the systems, as well as any accessory, buildings, plants, or vehicles that are an essential part of the system.

Intricacies of the Network

Water infrastructure comprises of a complex water supply network that has engineered hydrologic and hydraulic constituents which provide water distribution. The system in specific terms includes the drainage basin or the area of land where surface water from rain and thawing snow or ice congregates to a particular site, where the water body joins others such as a river, lake, reservoir, estuary or ocean. Also part of this network are the collection points above or below the ground where raw water accumulates, such as a lake, a river, or ground water from an underground aquifer. This water may then be transported using ground-level aqueducts, covered tunnels or underground water pipes to water purification facilities from which the treated water is further transferred with the help of underground water pipes. Water storage provisions in the form of water tanks, water towers or pressure vessels also form an important part of this system. The distribution of water to the final consumers—residential, industrial, commercial etc., is carried out by a network of pipes. The sewer system, not considered as a part of water supply system comprises of underground pipes and surface ditches in some developing countries. Reservoirs and canals that constitute the country's irrigation system along with pumping stations and gates that assist in controlling floods are also part of the larger water infrastructure.

All these constituents play their critical role in ensuring sustainable management of water resources and help in addressing the scarcity problems. Physical infrastructure is required to provide free flowing water supply and sanitation for the various consumer groups—domestic, agriculture and industry as well as for clearing and purifying wastewater. On the other hand the hydraulic infrastructure provides hydroelectric power and navigation. Further, several water infrastructure constituents are expected to complement the natural aquatic bionetworks in acclimatizing to a certain level of pollution and in dealing with droughts and floods in a better way as well.

An efficiently functioning water infrastructure is broadly regarded as reflection of and a facilitator to the country's economic growth. The constituents are quite advanced in most developed countries but their development in India and several other parts of Asia[1], however, is noticeably patchy.

Through the entire network of constituents discussed earlier, there is, therefore, an urgent necessity to plan and execute water infrastructure development on a faster track. For countries like India, that have clear cut priorities with regard to drinking water supply and sanitation for the growing domestic segment, the capacity enhancement has to be across the broader landscape to include industrial and commercial water supply, hydropower, wastewater treatment, flood control and irrigation. Apart from regulating natural resources like rivers and lakes to improve the provisions for water storage to meet the requirements sustainably, investment needs to be diverted towards creating non-natural basins which will be useful in several parts of our country where precipitation is erratic and there are continued dry spells.

In the developed world that includes some of the nations in Asia Pacific, a common thread is visible in how they manage the flow volumes and water quality in major rivers. Substantial investment in water infrastructure over the last few decades has helped these nations deliver quality water for domestic and commercial purposes, for irrigation and hydro energy and have also avoided contamination and disasters like flooding. Over these years, availability of skilled human capital to support the modern infrastructure has also increased.

Water Infrastructure Governance

In contrast to the aforesaid, the condition of water infrastructure in India is not really mature and is fast deteriorating in all aspects under this huge demand pressure from various consumers. The growing population and its inflow into already over-crowded urban centres are resulting in spiralling demand for freshwater. With many additional mouths to feed and with constantly falling yields agriculture sector needs more water for irrigation. Although service sector dominates in terms of contribution to GDP, India has a thriving manufacturing sector and especially for steel and chemicals production huge volumes of water is required. The challenge does not end here. The climate change experts closely analysing the water resources have already predicted rapid depletion in the proportions of Himalayan glaciers which will restrict water flows into the major rivers. The underground water table is plummeting drastically over the last two decades and is now well below 300 meters in northern parts of the nation. The challenge therefore is right here and is being compounded by the lack of attention and planned development.

Governance around investment in water infrastructure and the sector's overall management can help improve welfare for communities. Water Poverty Index[2] is an important tool and a cumulative index that explains lack of freshwater. The index comprises of five key components—resources, access, capacity, use, and environment. The development of water infrastructure should be seen as strongly linked to the long term upwardly correction of the human development index. Access to better water infrastructure, as the outcome of an extensive study conducted across Asia by Intizar Hussain[3] from International Water Management Institute, Sri Lanka, points out, has positive impact on crop productivity and farm incomes.

It is, however, not entirely true that the crisis is at the level of the resource's availability. The diminishing availability of surface and ground water does face the herculean task of catering to 16 percent of the world's population. But the problem is aggravated by the infrastructure for water supply and sanitation. There are reports that claim that there has been some improvement in the last 4-5 years, but it really depends on what we are setting as benchmarks when it comes to access to water sources. The claim that over 90 percent of urban population, and over 80 percent of rural population has access to water[4], needs further verification—is the supply 24×7? Is the water of potable quality? Some of the answers may not be all that convincing.

The abysmally low investment in developing water infrastructure results in the poor quality of water services, and coupled with poor cost recovery forms a depressing cycle. If the indifference continues in the unorganized water sector, the shortages that may not be obvious for now will become acute in the very near future. A large part of the Indian consumer population whether domestic or commercial, will suffer limited access to reliable water source unless there is a substantial focus and investment in developing water infrastructure. Needless to say—a mandate of this proportion cannot be managed by the government alone. The potential in the private sector needs to be further released to shoulder this emerging risk and provide the financial or non-financial resources and know-how.

Agricultural and Domestic Supply System

The maximum consumption of water in India is from the agriculture[5] and residential segments. Developing the infrastructure capabilities in these sectors will provide opportunities to domestic as well as international investors.

Over the last five decades, direct public investment of around INR 90,000 crore has been made to augment the irrigation infrastructure[6] of various types and scales. In order to keep up with the multiple users and usage, an annual investment of INR 20,000 crore was envisaged over the next two decades. The annual expansion however is happening at a much slower rate of about INR 7000 crore. Actual government outlay is declining over the last few years resulting in shrinkages of the areas under the major, medium and minor irrigation systems as witnessed in some major states. The general trend in South Asia as well as in India is to rely on ground water as the critical resource for irrigation water. In the present scenario, India has over 20 million modern water extraction structures. Almost a quarter of the farmer household has a tubewell; and almost 50 percent are buying the services from tubewell owners. Sustainability of this resource has been seriously threatened by major issues such as rapid depletion, brackishness and effluence. Although irrigated agriculture is of prime importance to support food security and key ecological assets, there is inadequate investment in the maintenance of irrigation water application and delivery systems. The decisions on irrigation infrastructure in India as well as in many other countries are surrounded by debates around other uses including recreation, environment, ecosystem functions, urban, or hydropower uses.

Starting with immediate effect, sustained efforts need to be made to restore the irrigation infrastructure and kick off all round institutional reforms in the water sector. The wider impact that is expected from the irrigation management systems will require this support that will enable them to play their part in water and food security. The irrigation sector needs to break away from the vicious cycle plaguing infrastructure, by empowering the stakeholders to conserve and deal with water resource management. The water policy has to take into account not only the restoration of the irrigation systems but also mandatory institutional development and its legal empowerment to identify user categories, for broadening the revenue base and enforce quantitative measurement of water supply, charging and collection.

In terms of growth, the Indian water infrastructure construction market is projected to expand at a compounded rate of 13.22 percent. It is an opportunity for private industries that supply water pumps and irrigation equipment. The Indian government has already announced thrust on supporting investment in water infrastructure in the current five-year plan. A total of INR 2,715 billion has been allocated for the construction of new and the development of exiting, irrigation water infrastructure.

The residential segment is the ensuing top consumer of water. In addition to the urban water supply utilities, bottled water and domestic water purifiers are the major constituents. Growing urban population, inflating city limits, limited ground/surface water, and increasing water demand are creating mandate for new water distribution and sewer collection networks. According to present estimates, approximately 63 percent of the urban population is connected through piped networks for water supply, and approximately 46 percent is connected through sewer networks for sewerage disposal. In order to improve living standards of urban areas, priority issues among municipal corporations and urban local bodies should be to provide 100 percent water and sewer coverage.

That said, achieving close to 100 percent water coverage implies demand for various products such as pipelines, valves, pumps and metering systems. The existing water and sewer networks are very old and need to be revamped in order to improve efficiency of the network. Further, contrasting with data even from peer emerging countries, the number of people connected to a house's water connection is very high in some cities and this adds to the water stress. The country also faces challenge in the form of relatively high instances of water theft and unaccounted for water in the water distribution network adding to the inefficiencies. Installation of new metering systems in order to minimize such inefficiencies becomes an imperative. Treatment and recycling of wastewater is an important area of overall water economy, which has remarkable scope in India. In a research report just a few years ago, Centre for Science and Environment discovered that Indian cities produce close to 15 billion litres of waste water on a daily basis[7] of which, as much as 30-40 percent of the domestic waste water is being discharged, without any adequate treatment. With no new sources of large volumes of water in sight, efficient usage and reusage of precious water resources could be one part of the answer to the problem of water shortages. For inspiration, India can look eastward at Singapore that has allocated hundreds of millions of dollars over the years, to create a comprehensive infrastructure that enables it to treat and reuse sewage.

SINGAPORE WATER INFRASTRUCTURE

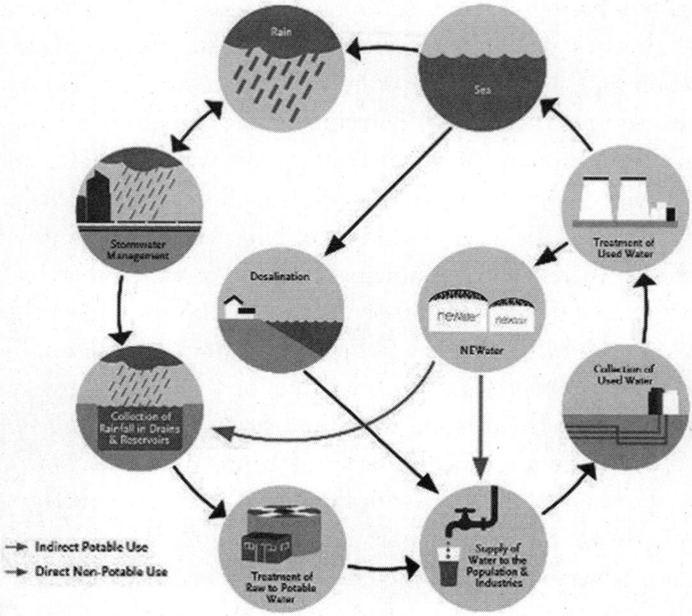

The major supply gap due to frequent water cuts and suspect quality of water supplied by the utilities has provided an opportunity for the private sector players to come in and develop the market of packaged water. The industry has seen an exponential growth in India, with the present size of the packaged water industry estimated to be in the region of INR 32 billion (or $583 million)[8]. Although the wide range of consumers is revolutionizing packaging, water in India is primarily sold in small pouches or in 1,000 ml bottles. The water purifier industry has also benefited from this gap and domestic installation in urban India has been on a consistent rise.

Touchstones Around the World

There are lessons to be learnt from the developed countries like the US[9], that have been investing billions of dollars over the years to build an extensive network of drinking water, wastewater and storm water infrastructure to provide the public with safe and clean water. Some of them have had the basic infrastructure for over 100 years but are now investing in the network of water treatment plants, distribution lines, sewer lines and storage facilities. Age and deterioration have taken a toll. Every year across the US, there are nearly 240,000 water main breaks. Further, some 75,000 sanitary sewer overflows annually discharging three to ten billion gallons of unprocessed wastewater, leading to various illnesses due to contact with contaminated water. The environmental researchers are trying to make a comprehensive assessment of the gap between the needs and spending on the water infrastructure in the US. This potential gap of around $500 million will compel other stakeholders into taking action to meet the challenges of the nation's aging water infrastructure.

Emerging countries like China and Brazil, which are rapidly industrializing have made significant investments in water infrastructure and trying to address the water risks. They may not have achieved complete infrastructure and institutional capacity to manage their water resources but are working towards providing widespread and dependable water services while optimizing sustainable growth.

The story in China[10] is somewhat similar to that of India where some of the thickly populated, agriculture focussed communities face acute water challenge. The need for strategically planned investment in developing the infrastructure is irrefutable and China has not been any less enlightened about its growing water crisis, as the policymakers crafted the first version of its National Water Law way back in 1988. The state's far-reaching as well as tactical plans to tackle the growing water shortage and quality issues include a strong commitment to development of water infrastructure sustainably. A lot of attention is being paid to the nuts and bolts of the nation's municipal and industrial water infrastructure, where substantial technical and operational advancements are required in water and wastewater management, supply and collection. The reasons driving the new approaches to more effective water infrastructure are the political will and the focussed investments that the Chinese government has been able to deploy.

All of these objectives are leading to new infrastructure, upgrading of the technology in existing assets so that more stringent levels of treatment can be reached, using new management techniques and replacing equipment to reach greater efficiency in conveyance. As for India, the progress as indicated in the recent plan, years may be limited but the situation appears to be changing slowly. It is obvious that the improvement and strategic construction of water infrastructure will not be as straightforward as simply building more infrastructures, upgrading equipment and increasing efficiency. That said, in the area of pollution control, China has seen significant progress, with current government figures placing the levels lower than the 2005 levels. China has reinforced its policy position through the joint declaration of the National Indigenous Innovation Product Accreditation Programme by the Ministry of Science and Technology (MOST), the National Development and Reform Commission (NDRC) and the Ministry of Finance (MOF). The programme gives specified predilection to a list of qualified enterprises in government purchasing and contracting, an amount that totals RMB 300 billion for water infrastructure.

In the long term, China holds the promise to be one of the global leaders in development of water infrastructure and technology. The recent policy imperative is very much backed up by an environmental as well as economic ultimatum. At present, China is still on its way to getting closer to that foresight. Conventional technology of limited scalability still dominates the water markets. Many of the Chinese cities grapple to reach 50 percent wastewater treatment and often have leaking distribution or collection systems. Nevertheless, with the level of local innovation, inclusion of foreign expertise and planned investment persistently on the upswing, China may well meet its goals for the long term.

The situation in another emerging market in the west, in Brazil, is of some remarkable achievements, slightly weakened by continued challenges. One of the key accomplishments for the nation is the 80 percent coverage. Close on the heels of this is the 78 percent access to sanitation network—again a massive increase in recent years. The central system that has without any interruption financed the water and sanitation infrastructure is one of the principal factors to which this achievement can be attributed to. The country's notable progress in advancing technical and financial innovation such as the condominium sewerage system and a government subsidy based on output for treated wastewater, has also played an essential part.

These advancements, however, have been challenged by Brazil's large population and the high urban population density—as much as 84 percent of the total population. As has been seen in several parts of emerging world, many of the poor urban occupants live in the slum dwellings without much access to safe water and healthy sanitation. To add other dimensions, geographically, the north east of Brazil is experiencing increased water scarcity, and the south east of Brazil is living through the water pollution menace. These demographics add to the water infrastructure quandary.

In spite of these challenges, the current scenario is being considered as promising both for the consumers and even investors. In a recent announcement, the national integration ministry, recently released $41.2 million[11] for water infrastructure projects involving the whole countryside under the country's growth acceleration plan called PAC. The announcement also included the plan for a major reversal project of the polluted Sao Francisco river and several water infrastructure projects for other states. Experts forecast that the total investment in water and the conforming sanitation infrastructure can reach $4.5 billion or approximately 0.7 percent of Brazil's GDP. This clearly provides a tremendous opportunity for international financial institutions and international water companies to take a closer look and further evaluate the potential. Even as the local commercial banks or the Brazilian stock market have provided the much needed capital, the international community has been showing keen interest as well. Interestingly, the Sao Paolo based water utility company Sabesp, is listed on the New York Stock Exchange.

Fostering Partnerships for Sustainable Infrastructure

The water infrastructure in India is estimated to hold business opportunities worth over $32 billion annually for the private sector. Given the capital intensive nature of this industry, private sector partnership or public private partnerships[12] will be successful business models to address mounting water demands and perpetual funding constraints faced by municipal corporations. These public bodies can decide to work with the private sector partners to not only bring in investments but to also introduce professional managerial expertise and world class technical resources. These partnership models can, therefore, provide a vehicle for highly efficient and widespread delivery, while retaining public ownership of the assets. With exceptionally high levels of government investment unlikely to continue in the long term, private sector investment in the developing water and waste management infrastructure as well as economically sustainable tariff structure for various user segments, should be increasingly acceptable. Traditionally, the relative lower rate of returns provided by water infrastructure projects, including water supply and sanitation schemes, has made the private sector more disinclined to invest in this sector than in other forms of infrastructure development. For many private sector players, entering the water industry will require a better understanding of the complex corporate and social risks involved. They must understand and participate in the overall water policy dialogue concerning pricing and regulation. A thoroughly well-structured and executed private sector involvement in the water industry can be very successful and it can deliver long term benefits not only to the government and the investors but also to the end users and the environment. There may be a requirement of a distinct publicly accountable body in the water sector that helps create an enabling environment, working closely with the private sector as well the communities.

Several organizations associated with the nation's water infrastructure system are possibly trying to work hard towards greater sustainability. This effort has been undermined by the low investments in scaling up and refurbishing the key constituents. And as the gap between the arrangements required immediately and availability of investment sources widens, the water infrastructure in the country will have to grapple with more of these issues.

India has this great opportunity to work with partners across the water and finance sectors to develop capabilities and devices that will ensure sustainable investment in water infrastructure. The multiple objectives of meeting the demand from various consumers, improving all round quality and avoiding recurrent disasters can all be achieved through robust infrastructure planning and orientation to best management practices. Effective planning is essential for water and wastewater systems to manage their operations and infrastructure and ensure the sustainability of the communities they serve. The evolving utilities management devices should clearly demonstrate how they are contributing towards improving efficiencies. Combined with the induction of better skilled water sector workforce, the system should also be able to promote sustainable usage and reduce wastage. Utilities that will continue to incorporate sustainability factors right into planning will benefit in the long term. The transparent and inclusive process deployed for project selection will help boost economic as well as environmental and social returns. Further, by way of consistently assessing the range of alternatives that addresses multistakeholder objectives, greater enhancement of financial as well as human capital associated with the project.

These efforts will be augmented by an equally vigorous asset management process that initiates the right level of investment in alternative technologies at appropriate times. Globally, water technologies have advanced and the framework should assess the latest innovations for solving the water infrastructure issues. In keeping with the precarious state that it is in, it is important that sustainable, future ready and cost-effective procedures are increasingly adopted. Further, very important for the sector is a more evolved dialogue and consensus around the pricing of water services and financing water infrastructure.

Even as the policymakers, public and private sector actors play their part in advancing the country's water infrastructure, communities can also collaborate to maintain healthy waters and contribute to sustainable management of resources. Participating in the execution of workable solutions to better manage storm water, storage, waste treatment and plugging the water leakages in their localities will have a very positive impact on sustaining the larger initiatives. At this stressful time when so much of our infrastructure is in need of accumulation, replacement or repair and limited financing arrangement to fund these projects are available, the time tested and inexpensive solutions can be of assistance in meeting several goals at the same time.

Case Studies

Nagpur PPP- For 24×7 Water Supply

Nagpur, often referred to as the second capital of Maharashtra, has a population of over 2.5 million spread over 217 sq km. In 2008, total water supply to the city was over 500 MLD per day through nearly 225 thousand connections. The distribution network consisted of over 2100 km of pipe line network. The study details out the goal of Nagpur Municipal Corporation (NMC) to achieve, 'water for all and 24×7 supplies with focus on safety, equity, and reliability'. The total sanction of funds, through JNNURM, before this project, was around INR 900 crore for the expansion of water supply system, and an additional INR 387 crore for rehabilitation of the distribution system. The Corporation has recovered the entire operation and maintenance (O & M cost, as well as the partial capital cost from this initiative through the fund. In order to achieve this, it had planned to adopt measures for effective billing and collection from consumers. A new tariff scheme was established so that a total O & M recovery was possible for the city water supply operation. It was the first instance in India that the entire O & M of the water service was outsourced to a private operator for such a lengthy duration.

A public private partnership (PPP) approach of financing was adopted to meet the mega objective of 24×7, uninterrupted water supply, as opposed to the 2-12 hours a day supply that was in place earlier. The capital expenditure included, was distributed to rehabilitate, repair, maintain and to provide appropriate refurbishment and replacement of water supply infrastructure. The project was conceived as an agreement with Orange City Water Pvt Ltd (a joint venture of Veolia Water, a global water giant and Vishwaraj Infrastructure, a local company with PPP projects in India). This PPP initiative in the municipal water sector was a first of this size in India.

The total expenditure was divided by the partners as follows—50 percent by JNNURM, 20 percent by the government of Maharashtra and 30 percent by the operator (private sector). The first five years of the programme was mainly to upgrade and rehabilitate the distribution network, with approximately 15 percent of the total expenditure. The contract included drinking water production, treatment, transport, storage and distribution to the end user.

Background

A couple of programmes to improve the water supply system had been adopted by the city over the years. In 2003 a Pench III, Water Loss Control Programme was commissioned for augmenting the city water supply by 120 MLD. The project included increase in capacity of raw water pumping, construction of water treatment facilities, water mains, elevated service reservoir and of distribution network. It was expected that by increasing the infrastructural capacity, there will be improvement in delivery. However, although the input volume was increased by 32 percent, water sales figures remained static. The entire project was converted into water loss in terms of both technical or real losses and commercial losses. With time, NMC had to face severe issues with financial losses increasing many fold and repayment of loan becoming a major share of NMC's budget. As a result, dissatisfaction pervaded amongst the stakeholders—citizens and people representatives—for not receiving enough water. The authorities had a difficult time explaining to the public and the media, the reasons behind continuation of water problems, when water supply to the city was more than 200 LPCD.

The Programme

NMC, then, undertook the Water Loss Control Programme (WLCP), to identify the reasons, and solutions, by taking advantage of the financial assistance programme of government of Maharashtra.

The WLCP made use of auditing techniques for leak detection and control. The programme in four phases is described below—

- *Phase one*: Water audit and leak detection study, assessment of economic optimum volume of water losses and performance indicators as done through water and energy audit study
- *Phase two*: Pilot study at Dharampeth zone for 15000 connection and leak detection in 53 test zones spread across the city
- *Phase three*: Full city intervention using apparent (commercial) and real (technical) loss reduction methods based on outcome of first two phases

- *Phase four*: Implementation of 24×7 water supply programme under long term performance based rehabilitate-finance, operations and maintenance contract by public private partnership mode for sustainability.

Water audit Phase-One was approved for ACA under JNNURM. While many components of the programme were completed, the remaining, as given below, were expected to have been completed by December 2012.

JNNURM Projects	Cost In INR crore	Outcome to Improve Supply Side Efficiency
Leak Detection	3.30	• Flow Meters from source to ESR • Reduction in BW –UFW to 2% from 23%
Water Audit	25.00	• 24×7 for 15000 connection by Sep'09 • Replacement of 20000 consumer meter • Reduction in UFW < 25% in Pilot
Energy Audit	28.79	• Replacement of All Pumps installed prior to Year 2000 with minimum efficiency of 70% • Reduction in system head by 12m for Pench-1 • Centralized monitoring system • Infrastructure for equitable distribution
Expansion of Network	43.49	• Connectivity to 572 layouts • Population over 2.0 lakhs will be covered in piped water network

JNNURM Projects	Cost In INR crore	Outcome to Improve Supply Side Efficiency
Kanhan Up gradation	87.65	• New Plant of 240 mld and discarding the old plant of 120 mld with Pumps
Pench-4 Part-1	211.58	• Replacement of Canal by 2300 mm pipe • Additional Raw water of 115 mld from reduction in water losses in canal
Pench-4 Part-2	79.74	• New WTP of 115 mld at Godhani • Trunk Main up to MBR
Pench-4 Part-3	87.98	• Service Reservoirs 25 nos. • Feeder Mains 35 Km
Pench-4 Part-4	174.00	• Distribution system of 225 Km

WLCP Phase Two: Pilot 24×7 Project and DMA's Leak Detection

NMC had undertaken the following projects to prepare the strategy for global water loss reduction programme

- Pilot project for 24×7 water supply to 15000 connections
- Test zone (DMA) for leak detection
- Maintenance and repairs record for reported leaks

Investigations carried out during the phase two enabled to gather valuable information regarding the existing situation of the system and highlight the main issues and weaknesses of the existing distribution system to be tackled, namely—

Technical or Real Losses: Major reasons for technical losses are:

- House service connection, as it accounted for 37 percent of reported leaks
- GI pipes were responsible for 44 percent of reported leaks
- Tertiary network (150 mm and below) were responsible for 92 percent of reported leaks
- The results of test zone suggested that most of the non-reported and background losses occured on HSC and the pipe connected with HSC
- The leaking pipes in the network were identified through complaint records for polluted water, frequency of leaks, and low pressure at specific locations

Other reasons for technical losses were:

- Evaporation and seepage at the Gorewada Tank
- WTP losses at Pench-II

Commercial or Apparent Losses: The major reasons for commercial losses were:

- Non-working meters and unmeasured consumption: NMC records showed over 206,000 recorded connections for water supply, and only 36,500 out of a total 154000 metered connections were shown working. The balance, over 62,000 connections, were billed with flat charges
- The accuracy of working meters was a major concern due to poor quality and ageing of meters. Study of 1333 working meters as per NMC records were checked and it was found that an average 18 percent was under consumption

- Illegal connections were a major concern, and in the test zone 17 percent connections were found to be illegal, of which 80 percent were in slum areas. Only 69 percent of the connections out of a total of 4133 connections in 53 test zones were issued regular bills
- The inefficient billing system and limited IT application lead to poor follow-up and collection efficiency. 13 percent consumers with legal connections in test zones were not receiving bills
- The non existence of a structured consumer service centre resulted in large number of disputed and unpaid bills
- Poor recovery mechanism lead to illegal adjustment and corrupt practices

The water losses (technical and commercial) had resulted into the following concerns for NMC—

- Inequitable water distribution
- Low level of pressure
- Significant number of interventions on the network per year and leakages
- High level of NRW
- Polluted water and water borne diseases
- Poor tariffs lead to poor cost recovery and inadequate budget for maintenance of assets. It resulted in deteriorating efficiencies

These results were analysed for preparation of the rehabilitation plan. Recommendations to improve the existing situation in terms of infrastructure and level of services were proposed in the rehabilitation and investment plan as given below—

WLCP Phase –Three: Rehabilitation for Entire City

Sr. No	Rehabilitation Plan	Unit	Qty.	Amount Estimated Cost (In Crores)
1	Inside Slum			
i)	Replacement of Pipeline inside Slum	RMT	175933	44.09
ii)	House Service Connection (inside Slum)	No.s	124274	
2	Outside Slum			
i)	Replacement of Pipeline of different diameter	RMT	253657	251.59
ii)	House Service Connection (Outside Slum)	No.s	206360	
3	Specials (Outside Slum & inside Slum)			15.88
4	Repairs of pipe & Specials			8.54
5	Electromagnetic Meter	No.s	100.00	7.64
6	Rehabilitation of ESRs	No.s	28.00	8.26
7	Rehabilitation of Pench II WTP	No.s	1.00	6.46
8	Replacement of Sluice Gates & Valves	No.s	28.00	3.59
9	Developing billing Software	No.s	1.00	2.00
10	Other Works			1.66
	Total			**349.71**
	Total BOQ (in Rs.) say			**350.00**
	Add @ 3% Contingencies			10.50
	Add @ 75% Project Preparation & Desgin Consultancy			2.63
	Add @ 3.5% Project Management Consultancy			12.25
	Grand Total Rs.			375.38
	Say Rs.			376.00

Tariff Revision

NMC had passed a resolution to adopt the tariff revision for full cost recovery as given below, and it was implemented in December 2008.

Category	Old Tariff /1000 ltr	New Tariff	Remark
Residential	INR 3-4	INR 5-15 as per telescopic consumption	Minimum charges replaced by monthly access charges (INR 56 for 15mm)
Semi Residential	NA	INR 8-18	Min. access charges are INR 100
Institutional	INR 12	INR 15-20	
Commercial-1	INR 20	INR 25-100	
Commercial-2	NA	INR 60-540	Mineral water + Cold drink
Slum	INR 25 PM	INR 30-80 PM	Kaccha, Concrete, Multifloor
Indexation	NA	Energy + Raw water	Directly passing to consumer as surcharge
Annual Revision	NA	5% to MC and 10% to 25% to GB	No need for approval from GOM

WLCP Phase – Four: O & M Strategy for Sustainability of Sector

The effectiveness of the investment plan and reforms depended on improved operational efficiencies and better accountability of service provided to consumers. Hence NMC had decided to transfer most of the risks of implementing the project and of operating the water supply system to a private operator under proper PPP arrangement based on the following principles—

- Ensure transparency in project management process
- NMC's control in providing water resources and in setting up the tariff and ownership of assets
- Achieving required operational efficiency assigned to private operator

The operator was handed the water supply system to operate and maintain on a BOT/ Lease basis.

Gujarat Model

Gujarat government has created an enabling business environment and a culture that promotes initiatives for development of water infrastructure. It is the mindset that governs the environmental behaviour, and until now, natural resources were taken for granted, as it was considered that it would last forever. This pitfall in mindset is well-identified by the Gujarat government and it is in the process of bringing change in thought processes, institutions and mechanisms. The government recognizes the need to be environment-friendly and adapt and mitigate climate change impacts, make minimal use of natural resources and gain maximum benefits out of them.

Attitude Matters

Gujarat's model of governance, in recent years, is characterized by an organized and holistic approach towards development of water resources. Gujarat has moved out of the traditional approach and undertaken initiatives to bring fundamental changes and aims at qualitative and quantitative leaps. The government has evolved and implemented technology innovation and far-fetching solutions to overcome the historic problem of water conservation. With the kind of initiatives, the once water-despaired state is now confident to say that there is absolutely no shortage of critical inputs like power and water in Gujarat. The state aims at maintaining this situation and continues to remain the preferred business destination. Recognizing that there will be an increasing need of water from all the sectors, the government is proactively making efforts of water management to avoid any kind of water crisis situation. Gujarat authorities have made hundreds and thousands of water harvesting structures, and also successfully implemented one of the world's biggest dams, the longest canal system and a state-wide water grid. Gujarat has initiated a movement away from water scarcity to water management. The government has a wide coverage to the water distribution network, and it aspires to cover even the remaining water network with piped water supply. To manage water in agriculture, the government is determined to support and enhance scientific agro-practices and agro-processes.

A large number of infrastructure projects are being undertaken by the centre as well as within the state, because of the increasing housing and transportation demands, accompanied by development. To support the growing needs within the state, new projects and capacity building of cities and municipal bodies are being undertaken, and the state government is highly receptive to these initiatives. With a wide spread focus, initiatives are planned equally for the small and medium towns within the state. According to the latest data, Gujarat ranks fourth in the category of getting projects completed. In the year 2011, the government in centre awarded Gujarat for showcasing best water management among all states.

To take an overview of the water sector and improvements therein, in Gujarat 20 percent area had 71percent water resources, and remaining 80 percent area had merely 29 percent water resources. Therefore, managing water resources and improving water access was a big challenge. The government, backed by a strong political will, carried out the exhaustive task and turned crisis into opportunity.

The WASMO Project

For improving access to rural areas a 'Special Purpose Vehicle' was created—WASMO (Water and Sanitation Management Organization). This project has yielded tangible and intangible outcomes of empowering the rural community in water supply and sanitation. The key element of the model is involving rural citizens in the decision making process of water supply in a decentralized format laid as self-sustained villages. The involvement is in the form of Pani-Samiti, a sub-committee of Gram Panchayat, and sustainability is ensured by inclusivity and adequate representation to women and marginalized groups in this committee. In order to develop a sense of ownership, the Pani-Samitis are directed to bear 10 percent of the capital cost of required infrastructure. Institutionalization of women has lead to positive impact also in terms of sanitation, health, control of water borne diseases and reduction in drudgery. Further, to ensure safe drinking water, another supporting initiative—'Rural Drinking Water and Sanitation Programme' was implemented. Partnerships with organizations like UNICEF and the state agency has helped the initiatives and implementations in a big way. The problem of water and sanitation is as much with public awareness as is it with the infrastructure facilities. To support the infrastructural and governance improvements, water quality teams were setup in different villages. These teams comprised of members from Pani-Samitis, self-help groups, students and anganwadi teachers, with the main function of increasing awareness about safe drinking water and personal hygiene practices. The baseline ideology of the initiative is to inculcate appropriate water and sanitation practices in the cultural and value system of the community. Indeed the initiative received world recognition with an honour of United Nations Public Service Award—2009. The credit is to the idea of fostering participatory decision-making and decentralization through a social process. The success is a classic example of sustainability and results of adopting all its elements viz. technology, ownership, community inclusion, consideration to women and marginalized awareness, and surveillance. The results not only show through awards but also statistical evidence. After implementation of WASMO, 72 percent households had a tap connection against the national average of 26.6 percent. The quality of life in rural areas has improved, with rate of women affected with poor living condition reduced to 7.7 percent from 53.8 percent. The initiative has substantially reduced the social issues relating to water, such as long distance hauls for water, conflicts in water distribution, dependence on external agencies and tankers for water.

Water Management and Conservation

The various water management and conservation initiatives have significantly improved the ground water level in the state, with 90 percent watersheds reporting more than one meter water level increase. Over 5.5 lakh water structures like check dams, bandharas, and farm ponds have been created in last 10 years across the state. Around 4.5 lakh hectare land is covered under micro-irrigation schemes. New projects on water harvesting, water harnessing and water management are being taken up at a very fast rate. One of the novel 'model-villages' in the country is in Gujarat, where water initiatives actually truncated the crop pattern, permitting farmers to cultivate a cash crop, and meet for its water requirement. The Gujarat State Watershed Management Agency, a nodal state agency, has implemented Integrated Watershed Management programme. Under the programme the natural resources like soil, vegetative cover and water are being harnessed and conserved with an objective of restoring ecological balance. The underpinning aim of this ecological restoration is water. The outcome of the ecological restoration is reduced soil erosion, increased infiltration and vegetation and thereby improved ground water recharge. To ensure accuracy and success, use of right technology and methods are deployed from the very first planning stage. Use of geo-spatial data for depicting baseline status about land, water, vegetation, village locations, analysis that includes socioeconomic considerations, land water treatment of area are the broad features. With increased ground water table, reduced soil loss, increased agricultural productivity, change in cultivation pattern and increased labour employability, there is enough statistical evidence to prove improvements. The Gujarat model is appreciated worldwide and high-level delegations are visiting the place to study water management in adverse conditions and seek technical guidance.

Plumbing the Gaps

'We build too many walls and not enough bridges.'

—Issac Newton

The demand for water investment in India is nothing short of staggering, which means the scope for investments in water is diverse and wide ranging. Just as India's water challenges are disparate and diverse, so are the investment opportunities. The nation's investment needs to cut across various sectors of the economy—agriculture, industry, transportation, tourism, healthcare, and energy, to name just a few. They touch every sector of India's emerging 'green economy'. Meanwhile, the water challenge before India must be met by both the private and public sector—such is the importance of water to India's future. A public private consortium is needed. National and regional coordination is also an imperative, as is closer cooperation on a regional and global scale. Key municipalities like Mumbai, Delhi, and Kolkata—among the largest cities in the country and the world—should also be actively involved in mapping out India's water future.

The investment opportunities involving water run the gamut, ranging from waste water treatment, to desalination processing, to hydropower creation, to the deployment of watersheds, and flood prevention policies. According to research from Merrill Lynch, the water industry could be worth $1 trillion by 2020. A large share of this growth will occur in water-scarce Asia, India included.

Infrastructure investment is one key area where more capital investment is imperative. In order for India to better manage its water capabilities and needs, the nation must invest more capital in the construction of water storage units, municipal dams, water treatment facilities, flood protection schemes, and ground water wells, and pumps, to name just a few areas.

Water storage and treatment will be critical to India's agricultural and industrial sectors, with these industries and many related industries already confronting water stress points that could hamper future growth, and impede the nation's development. In addition, municipal and residential water use is set to explode as India's pace of urbanization continues and the demand for clean water among urban residents soars. Capital expenditures on pumps, valves, pipelines, tubing, and related activities should be poised to explode over the next few years. Why?—Either India meets these challenges and makes the necessary capital expenditures over the next few years, or the economy faces stagnation, slow growth, unacceptable levels of grinding poverty, and political unrest. It is that simple.

Not only does India need to build its water infrastructure, but future capital expenditures on water have to be done in the context of other water-usage and related industries. For instance, in the area of waste-water treatment alone, the number of industries affected and involved include utilities, oil and gas companies, mining, petrochemicals, energy, breweries, and many other industries that are very water-intensive. The textile and pharmaceuticals industries are two more sectors where water treatment is key, with many firms in these industries already having in place established treatment systems. Investment plans and outlays must acknowledge the interconnectedness of all of the above. Investment should not be done on a piece-meal basis or in an adhoc manner but in a coordinated and cross-sector basis.

The challenge is to make the existing water supply more acceptable to desired end-use purposes, such as drinking water, usage or reusage by industry, in irrigation, and other activities. The key is this—India's investment opportunities are not only sector-specific but also cut across many sectors and industries, creating a huge multiplier effect in terms of need for capital and potential job growth.

In agriculture, capital needs are pivotal for irrigation systems, drought-resistant seeds and crops, the use of smart-metering and water monitoring facilities. Each one of these tasks will require significant levels of capital investment in the years ahead. Given how important the agricultural sector is to the Indian economy, based on the livelihood it creates for millions of Indians who toil on its fields, smarter and better irrigation devices are required. The potential for investment in this sector is massive.

Regarding irrigation, one additional note is worth mentioning and that is the expanding investment opportunities in micro-irrigation, where these techniques are only now beginning to be adopted. The spread of irrigation technology—including micro-irrigation, trickle-irrigation, daily-flow-irrigation, drop-irrigation, sub-irrigated-planter-irrigation—all of these applications are suitable for India and represent future potential investment opportunities.

And then there are the costs that go beyond building a modern-day infrastructure. Of critical importance are—developing and modernizing maintenance facilities, water storage and monitoring units, and pollution control stations against an expanding backdrop of volatility, global climate change, rising urbanization, and increasing residential population.

Given India's dependence on the monsoon rains, the effects of climate change are of particular concern since it can exacerbate the supply-side of the water equation. For India, this means more capital investment for flood prevention, water storage, water distribution and transmission, and sea walls in many parts of the country. It also should entail more investment on water retention methods and techniques, and a rainwater harvesting infrastructure.

And then there is the effect of global climate change on large glaciers of the Himalayas, which are of critical importance to both India and China. If there is a true common denominator that links India with China, it is the need for reliable water sources—neither nation will live up to its much hyped economic potential without a secure supply of water. Hence, the urgency and necessity that two of Asia's largest economies work in tandem to tap the natural bounty of the Himalayas. Neither India, nor China, should exploit resources independently, but in tandem and with cooperation.

India's energy future is also tied to the availability of water. As India builds its underdeveloped and overwhelmed energy sector, more and more pressure will be placed on the nation's water infrastructure. Oil and gas exploration and refining, coal and iron ore mining, energy pipelines—these and other critical projects that will require massive injections of capital over the decade, and should be completed with an eye towards India's underlying water needs and infrastructure. Water treatment for utilities will also soar in the years ahead, with both coal and oil-fired utility plants being huge consumers of water. The nation's energy and water infrastructure should be built jointly, with a judicious eye on just how much water will be required to drive India's energy industry.

In the end, India's growing water stress, coupled with the water-energy-food nexus, underscores just how important water is to India and the wide-ranging effects if the nation's water demands are not met. The stakeholders cut across multiple industries, including power generation, where water usage is critical in the cooling process, producing steam, condensing and processing waste, removing impurities, and transporting fuel through pipelines. In the oil and gas sector, water use is more indirect, but nevertheless, still very vital. Shale gas, in particular, is very water-intensive, with the end result in many cases being highly contaminated water that must be treated or disposed off safely. In the mining sector, meanwhile, water is a key input for the processing system, with the crushing processes needing a reliable and large supply of water. Other private sector stakeholders include food and beverage companies, whose operations will come to a grinding halt, without a clean supply of water.

Investment Prospects in Desalination

Desalination is the process whereby unusable water becomes useable, with the extraction of saltwater at the core of the process. Effectively and efficiently making unusable water useable remains a key technological challenge, although there has been a surge in the installation of new seawater desalination plants. Indeed, according to research from Merrill Lynch, the global market for it is presently, roughly $6 billion, but expected to grow to $17 billion by 2017 and $25 billion by 2025. As the report notes, desalination is now standard procedure in 150 nations, India included. Yet only 1 percent of the world's population is currently dependent on desalinated water to meet their daily means, portending a huge upside in future decades.

Among the global leaders in this field are the Gulf States, Australia, Central Asia and US. The largest markets for desalination, according to Merrill Lynch, include India, as well as China, Saudi Arabia, UAE, Israel, Spain, and the Caribbean. Suffice to say that the market for desalination in India is poised for significant growth in the decade, underpinned by India's ever mounting water challenge.

That said, finding the capital for such investments will remain a key challenge for India. Investors, therefore, should be very mindful of India's massive capital needs for a modern-day water infrastructure on the one hand, versus the prevailing climate of scarce capital on the other hand. As the United Nations notes, the developing nations need to invest nearly $75 billion a year to meet water sector demands. However, capital remains precious and in limited supply, five years hence the global financial crisis. India is no exception to this trend. The financial burdens of water are tremendous to the country.

Against this backdrop, the United Nation recommends the following for many developing nations—

'Narrowing the financing gap involves a four-step reform agenda. First, service providers must reduce a range of inefficiencies to increase revenues and lower costs. Second, providers should tap available public funding (including ODA and subsidies), while governments improve the efficiency and efficacy of funds. Third, having improved profitability and the quality of the service they provide, utilities can begin raising tariffs to reflect the real costs of service. Finally, once they have the necessary political will and institutional capacity, providers should form public private partnerships and apply for commercial loans.'

Further commenting on the need for capital financing, the United Nations notes—

'Closing the financing gap in the water sector requires the application of a range of instruments including higher collection rates, more efficient service provision with lower costs, more targeted subsidies, and higher user charges. It is likely to be a longer-term process in which the approximate mix of instruments will change over time. Such efficiencies, even in the absence of full cost recovery, will improve the ability of utilities to adapt to future risk, and will make them less dependent on external funding.'

The Rise of Water Utilities

The global water utilities industry is valued at roughly $725 billion but poised to continue expanding over the next decade. Presently, the market is concentrated in the developed nations, notably the US, which accounts for roughly 40 percent of the global industry. Among the developing nations, China represents the largest market, being ahead of India when it comes to deploying water utilities to help regulate and manage the nation's massive water needs.

But this is one activity India should actively pursue with both local and foreign companies, whereby water utilities would serve as a partner to the government in managing a more efficient and modern water infrastructure. There is massive potential for this type of activity in India but the initiative for more water utilities has to come from the national government.

India's Water Investment in 'Soft' Infrastructure

While India will need to spend billions of dollars on its 'hard' infrastructure—or capital investment need in pumps, values, levees, pipelines, plants, waste water treatment facilities, etc—the nation needs to commit billions more on 'soft' infrastructure spending. This encompasses a diverse set of industries and companies that supply equipment, analytical instruments, services, software networking services, and related activities to help run India's water infrastructure in general.

In this space, think of firms that test and measure water safety or provide environmental testing applications for analysing water for chemical and biological contaminants. Air monitoring for particulates, pollutants and greenhouse gases also falls underneath this category. Firms that provide soil and sediment analysis, and solid waste analysis would be included in this group. Another investable cohort—firms that provide households with water management tools and techniques, or products like high-efficiency showers, faucets, toilets, and other appliances that would help monitor and regulate water use in India's large cities and newer residential developments.

It is a diverse cohort but one nevertheless poised to grow robustly as India embarks on a more coherent and large scale investment push to meet its water challenges in the twenty-first century.

Corporate Leaders

On a global basis, who are the global leaders in the water investment complex? Which companies are either operating in India today or likely to have an in-country presence in the future?

The following companies represent a sampling of water investment leaders as per their respective industries, and serve as an outline of some of the world's most exposed and involved companies in meeting the global water challenge of today. The list is by no means complete.

They include the following—

Most exposed water treatment companies—Alfa Laval (Sweden), ALS Limited (Australia), BASF (Germany), Bureau Veritas (France), China Everbright (China), Danone (France), Doosan Heavy Industries (South Korea), Dow Chemical (the United States), Dupont (the United States), Israel Chemicals (Israel), Pall (the United States), Nitto Denko Corporation (Japan) and Thermo Fisher (the United States).

Water management companies—BASF (Germany), Deere (the United States), Itron (UK), Dupont (the United States), Melrose (the United States) and Monsanto (the United States).

Water infrastructure and supply solutions: KSB AG (Germany), Kubota (Japan), Beijing Enterprises (Hong Kong), Downer EDI (Australia), URS Corp. (United States), Aguas Andinas (Chile), Manila Water (the Philippines), Severn Trent (UK), and United Utilities (UK).

The firms just mentioned, and many others like them should be actively pursued as partners to the Indian government. The nation's water needs are simply staggering and too massive for the government alone to finance. Hence the national government should make foreign direct investment in India's water infrastructure a strategic economic priority and actively seek foreign investment from world class multinationals. The latter not only have the capital that is badly needed by the nation, multinationals also bring a bundle of growth benefits, ranging from technology transfers, managerial skills, distribution channels, best in class business practices and related activities.

India's water infrastructure cannot be developed and built without the capital, technology and other attributes of western multinationals. The latter should be a strategic target of India's foreign direct investment strategy.

Water as a Source of Stability and Economic Cooperation

A final investment that India should consider lies in so-called 'hydro-diplomacy'. By this we mean the following—as a nation critically dependent on water for its future growth and prosperity, India should take the lead in promoting more water cooperation in the world's top international organizations. This means giving more weight to water diplomacy in such institutions as the United Nations, the Asian Development Bank and the World Trade Organization, which can arbitrate water disputes between its member states when the disagreements are commercial in nature.

This tack would help raise the level of cooperation between India and China, and the nations's other neighbours, as well as, create more synergies with various multinational institutions with the skills and capabilities to help India overcome its massive water deficits. India should also look to water leaders among the developed nations, notably the US and a host of European states that have been at the forefront of developing clean, secure and reliable sources of water.

Given how important water is to India's future, it should be a champion of 'hydro-diplomacy', a stance that would signal to many potential investors—at home and abroad—that the nation is very serious about tackling its water difficulties.

The bottom line is that India's water investment opportunities are almost limitless—billions of dollars are needed to build the nation's 'hard' and 'soft' water infrastructure. The downside is that most of this investment is capital-intensive; a great deal of capital is required in a world today where capital investment in large projects remains out of favour with investors. In addition, the rate of return on investment for investors also remain unclear—a critical variable that requires more transparent pricing and government policies. Without these variables, firms will remain reluctant to sink their capital into any Indian projects.

For investors, India remains a very seductive market for investment. But investors must realize that many of these investments will remain underfunded in the near-term, and that when it comes to return on investment, something that is too good to be true is probably just that—too good to be true.

Charting the Flows

'For tomorrow belongs to the people who prepare for it today.'

—AFRICAN PROVERB

India's water crisis will only intensify if the government does not take concrete action to better manage existing and future water reserves. The nation's water challenges not only reflect a mismatch between supply and demand; it also mirrors the effects of weak policies and poor management on the part of the Indian government at all levels.

India's water challenges require leadership at the federal and local level, as well as leadership from the private sector and the special interests of agriculture, energy, transportation and a host of other sectors. In addition, meeting future water needs will require significant amounts of capital from both the public and private sectors; capital will also be needed from some key multilateral institutions like the Asian Development Bank and World Bank.

Yet another requirement lies with India learning from the rest of the world. The water challenge is global and hence various global responses have emerged over the past few years, with some best practices better than others when meeting the challenges of water scarcity. We outline some required policies (or needs) and some best practices below.

The Need for More Information

Best practices regarding water management begins with more information and data usage of water, allowing for more informed decision-making among policymakers. For much of the past, information about water usage was relatively primitive and underdeveloped since many governments around the world viewed water as an infinite resource—something that was always in abundance and plentiful for all.

However, as the following chapters have outlined, India's water supply is neither finite nor plentiful. Water management is now a strategic imperative for the country, which means the public sector requires better data collection and management about the nation's underlying water supplies and future demand, with greater knowledge by region, state and municipality critical to crafting long term strategic plans that benefit local users and fit into national goals and objectives. Because India's water supply is so variable, or subject to numerous uncertain metrics, the process of analysing incoming data has to be adjusted to a high degree of variability. The more information at the disposal of government and private sector leaders, the greater the odds of effectively managing water risks and uncertainties.

As the United Nations notes—

'One of the most direct ways of reducing uncertainty is to generate new knowledge or understanding of conditions governing water availability and quality in the present and in the future. Data collection, analytical capacity and predictive ability are all required to reduce uncertainty and therefore to facilitate decision-making about allocations, uses, mobilization and treatment. While the risk to water is not reduced, it is better understood.'

Also noted by the United Nations seminal World Water Development Report—

'Water demand management calls for a good understanding of who is making the demands on water and how much they are demanding. This requires considerable knowledge and information, without which policies may target suboptimal sectors or users, thereby having a negligible impact on future risks. Improvements in resource monitoring and databases of water use information are therefore required.'

And one more missive from the United Nations on the importance of information and data—

'Generating data for policy-makers and managers (observations, analysis, modeling, scenario-building) helps to inform decision-makers and reduce uncertainty. Effective governance in areas such as environmental controls, ground water monitoring and abstraction licensing, and monitoring and policing of pollution, can reduce the risk of over-exploitation of water resources and catastrophic surface water pollution and irreversible contamination of aquifers.'

The Need for Greater Education

Most consumers of water believe the commodity is infinite, whose price should reflect its abundance. Nothing, however, could be further from the truth. Water is one of the most mispriced commodities in the world, a trend that still prevails in the developed and developing economies alike.

Governments around the world, India included, must begin to educate all water consumers of the value and scarcity of the 'blue oil'. Better education of end users will help promote water efficiencies and conservation efforts, both which should be encouraged by reduced government subsidies for water usage.

The value of water is under-appreciated and therefore undervalued in the minds of consumers, a situation that can only be amended by more aggressive educational and awareness programmes from the government. It is imperative that the Indian government make its 1.2 billion people stakeholders in India's water challenge—asking, rallying and encouraging its population to understand and recognize the strategic value of water and its multiple touch points across the entire economy.

And beyond the general population, India's water leaders or key managers of water need to take a more proactive role and approach in informing decisionmakers in all sectors of the economy of the strategic value of water and how the entire foundation of the economy rests on one simple commodity: water. Without it, nothing grows or prospers—future economic growth will be retarded. In this endeavour, the government should enlist the private sector, NGOs and other key stakeholders.

The Need for Better Pricing

In many instances, the price of water is either too low or insufficient to create the right incentives for additional capital investment. Water is mispriced, with municipalities around the world, India included, not charging enough for water even to meet basic operational and maintenance costs. Contrary to many beliefs in India, water is not a gift from God, but an increasingly expensive economic input. Accordingly, the price of water needs to reflect the full value of water.

On this basis, the United Nations categorizes the following as the value of water—

Direct use values: the direct uses of water resources for consumption include inputs for agriculture, manufacturing, and domestic households. Non-consumption uses include hydro-electricity generation, recreation, navigation and cultural activities.

Indirect use vales: the indirect environmental services provided by water include water assimilation and the protection of habitats, biodiversity, and hydrological functions.

Option values: these refer to the value of having the option to use water directly or indirectly in the future.

Non-use values: these include water's bequest value (passing on this natural resource to future generations) and the intrinsic value of water and water ecosystems, including biodiversity, the value people place simply on knowing that a wild river, for example, exits.

Given all the above, which illustrates the multiple uses of water, water needs to be better valued, which would lead to better pricing.

Some nations are finally realizing and acting on this dynamic. Portugal now taxes major water users in agriculture and industry. China has also increased the cost of water for larger users and paying more attention to recycled water use. Singapore has also been out in front of many nations when it comes to pricing water more towards the actual value of the commodity.

Why are better pricing models needed? Because higher water prices would raise incentives for more efficient water use, and bolster the amount of investment capital needed to upgrade and modernize the water infrastructure of India. More revenue equates to more spending on technology and on India's 'hard' and 'soft' infrastructure.

When the opposite is the case—or when water prices remain low and the value of water is mispriced—a multitude of negatives transpire. The low value of water, for instance, results in lower water revenues, with in turn leads to inadequate water maintenance, operational inefficiencies, and inadequate investment. All of the above results in infrastructure degradation, rising pollution, deteriorating health conditions, low service quality and a host of other expensive challenges. There is a snowballing effect to mispricing, incorrectly valuing water in other words.

In India's case, more realistic pricing is urgent given not only because of the capital needs of India's water infrastructure. More market-driven pricing is also needed since at the moment, India has some of the lowest water tariffs in the world relative to other major countries. For instance, according to Global Water Intelligence (GWI), the combined average tariff on water and waste water usage in India was just \$0.15 per m3. That is just a fraction of the average water and waste water tariff for the 310 cities in the 2012 GWI Water Tariff Survey ($1.98m3).

India's combined tariff has increased over the past few years but still remains woefully low, denying the government of critical funding to develop a twenty-first century water system and network. To this point, India's combined water tariff of $0.15 in 2012 was well below China's ($0.49), Turkey ($1.63), the Czech Republic ($3.25) and Germany ($5.51). On the positive side, and in a bid to raise some badly needed capital, water and waste water tariffs rose 51.1 percent in Hyderabad between July 2011 and July 2012. Meanwhile, Aurangabad raised it tariffs by nearly 39 percent over the same period.

The Need for Adaptive Policies, Management and Governance

Australia, Israel, Singapore and China to a degree are governments that have been relatively successful in managing and adapting to multiple and varying water challenges. China's Hubei Province is cited by the United Nations for its wetland restoration programme, which not only helped restore the wetlands and contributed to flood prevention, but also restored many lakes and floodplains that have increased the biodiversity and water quality of various waterways.

Australia and Israel have been successful in controlling and limiting the deleterious effects of droughts; in both nations, water is treated as a key foreign security problem. The Philippines, like China, has made significant strides in wetland restoration, as has South Africa, where the Mondi Wetlands programme ranks as one of the most successful wetland conservation programmes in South Africa.

For India, a key challenge is for greater coordination and cooperation at three levels: the central government, member states and local cities and communities, overlaid with more cooperation with various national and state agencies that are all dependent on sufficient water supplies. Regular consultation with the private sector should also be part of this process.

The Need for Water Reuse

Water reuse must become part of India's water portfolio. Recent improvements to treating water should be considered by India given the country's soaring demand for water. Agreed, treating municipal wastewater and reusing it for drinking water, irrigation, industry and other applications hardly sound appealing, with many psychological barriers in place today that will be hard to overcome. However, advances in technology and design could significantly boost the total supply of water available in India through water reuse techniques. In addition, in all likelihood, the possible health risks to exposure to chemical contaminants and disease-causing microbes from waste water reuse do not exceed, and in some cases may be significantly lower than, risks of existing water supplies.

For best practices on devising strategies for waste water reuse, India should study the example set by such emerging markets as Kuwait, Israel, Singapore and Saudi Arabia, many of which are global leaders when it comes to per capita water reuse. The US reuses the largest volume of treated waste water, although on a per capita water reuse basis, the leaders are Kuwait, Israel, Qatar, Singapore and Cyprus—where water reuse was greater than 10 percent of total water extracted. In bone-dry Israel, roughly 75 percent of waste water is reused, with most of this supply going for agricultural purposes.

The Need for Water Trading

Given India's critical water needs, all options should be pursued, even options that might involve water trading. Chile, South Africa and Spain are all involved in water trading schemes, as does Australia.

The problem lies with the fact that water is heavy and energy-intensive. In addition, as one of the world's most strategic commodities, there is, not surprisingly, a great deal of resistance to the idea of trading water. In the US, the Great Lakes region has established laws and regulations that ban water trading. Nevertheless, water-rich Russia is in favour of the idea, while Singapore regularly imports water from neighbouring Malaysia.

What about the future potential of waterborne water trade? That depends on the cost of transporting water and the availability of the supply of water. New Zealand has inquired about the safety of moving large amounts of water by marine vessels, but has not emerged as a potential supplier yet. In the future, however, the water stressed areas of the Middle East, Africa and India may be at the receiving end of freshwater from New Zealand. Depending on the severity of the emerging global water crisis, international bulk shipping in water could take hold.

Need to Enlist the Help of India's and the World's Top Corporations

Only a handful of companies have fully grasped the strategic challenge of India's water threat. Many multinationals operating in India understand the risks, and have worked closely with the Indian government to secure a more predictable water supply. Coco Cola, for instance, falls under this category.

But more needs to be done. It is imperative that the Indian government and the private sector partner and strategize together when formulating plans and policies for India's water solutions. The latter should not only be leveraged as a source of capital but also a key source of technology, first-hand experience and source of best business practices that have been successful in other parts of the world. For its part, the Indian government needs to mobilize its resources and talent at the national, state and local level so as to effectively partner with the private sector. It is incumbent on the government to clearly lay out a path and plan that allows private sector participation.

The Need for Financial and Economic Incentives

Creating a water infrastructure is not cheap—indeed, the costs are rather high for any nation. Even in good times, investment capital for water is challenging, and even more so given the financial crisis-cum-global recession of 2008-09.

As the United Nations notes—

'Rising commercial finance for water has become more difficult due to the global financial situation since 2007, which has discouraged new private interest in water infrastructure projects, and had also unsettled partners in existing private public partnership (PPP) ventures. Early in 2009, the IFC reported that $200 billion of PSP projects had been postponed or had become "at risk", 15-20 percent of which were in water supply and sanitation. The financial climate affecting both the supply of risk capital (e.g. equity) and loan capital to finance these concessions deals, since liquidity has become scarce, and the problems of international banks have had repercussions on local banks too. Many innovative deals, developed with technical assistance and risk-sharing from donor agencies, are at risk.'

Suffice to say, the global financial crisis, while roughly five years in the past, has left deep scars on the investment community and their willingness to deploy capital when it comes to large capital projects. Public private partnerships remain key and have begun to revive in certain parts of the world. However, with many governments under fiscal and financial pressures, India included, and with many corporations hoarding their cash and reluctant to invest, funding for many PPP projects is lagging or not as forthcoming as needed.

Capital for India's water infrastructure remains challenging, in other words. Potential sources of financing can come from the government directly, via joint public private sector financing, private sector funding only, or by pursuing other measures that would raise revenue. The latter might include raising the cost for water and better rate of water collection. Regarding the latter, in Africa alone, some $500 million annually is under-collected or not paid, leaving governments struggling to find new sources of capital. Non-payments from public sector agencies is a key issue, and one that if reconciled would create a more predictable stream of revenue for future investment.

Another potential source: multilateral financing and aid from the World Bank or Asian Development Bank (ADB).

Underscoring the challenge of funding water, the United Nations notes the following—

'One basic reason why water does not receive sufficient funding is that its scope if often viewed too narrowly, whereas in reality it underpins a wide range of economic sectors, all of which would be threatened by water scarcity, pollution or pressures from the other drivers described in this report. Hence, a precondition of adequate financing for water is a full appreciation of the social and economic purposes that it serves.'

In general, India needs to consider a gradual elimination of its subsidies on water and electricity, and consider incentives to build out the nation's capacity via joint public and private sector projects. India's cost associated with a twenty-first century water infrastructure is just too staggering for the government alone to finance.

The Need For More Water R&D

In India and around the world, there is an urgent need for more water research and development and the encouragement of the international transfer of such knowledge. Pollution control, wastewater collection, treatment of reused water, emission mitigation, irrigation, hydropower, wetlands management—these and other activities are highly research and technology intensive and typically pay large dividends to both providers and end-users.

There is considerable scope for more R&D funding for these activities, but the Indian government, working in conjunction with the private sector, must take the lead in driving this process. The national budget should clearly outline how much funds are to be allocated to water R&D, thereby signalling to the private sector where commercial opportunities lie.

The Need for Regional Cooperation

In many cases, water knows no boundaries, often travelling across various external boundaries. This makes regional cooperation, or state-to-state management of water, a key mandate of many governments. Given the vital importance of water to all forms of economic activity, it is little wonder that in some parts of the world, water scarcity has become a tension point among nations.

Cross border management of water will increasingly become a key challenge for many nations, India included. One example of this arrangement comes from Southeast Asia, where in 1995, Cambodia, Laos, Thailand and Vietnam established the Mekong River Commission, which helps manage the river's minimum stream flow during the dry season. The agreement has expanded to include greater cross border management of not only water flow but also the entire ecosystem of the Mekong River, all with an eye towards creating a sustainable and predictable level of water for cross border usage.

In general, water management and planning should not be done in a 'silo' or irrespective of other regional players. Part of India's water challenge is very much cross border in nature, which suggests that the Indian government needs to be more proactive in managing its water resources with its neighbours, notably China. The two parties should work towards sharing more information with each other, with an eye towards developing joint strategic solutions to their water problems. In this respect, India should also be quite active and engage with the technical and financial resources of the World Bank and ADB.

Conclusion

Across Asia, nearly 500 million people lack access to improved water resources, while nearly 2 billion have no access to adequate sanitation facilities. This is a crisis that requires immediate attention from heads of state around the region, lest the so-called Asian economic miracle fades and the region slips backwards, with millions of people stuck in poverty and abject misery.

Meeting the water challenge requires that India address collectively and singular the pressing issues before it, including rapid urbanization, industrialization, intensive agricultural use, soaring pollution levels, extreme weather patterns and the rise of a middle class that demands and expects clean water in their daily lives. Without clean water, India's economy is bound to underperform, if not stall in the years ahead. It is that simple.

The same fate confronts Asia's other giant—China. This mutual need could help pave the way for more Indian-Chinese cooperation in the future, with water scarcity, possibly, becoming a uniting and cooperative issue that results in more bilateral cooperation between the two parties.

As this chapter underscores, there are great number of needs for India to address in the future. How well the government addresses these issues in the future will determine the overall fate of the country.

Notes and References

Rising Premium on Blue Gold

1. Global Water Security, Intelligence Community Assessment, February 2, 2012

2. Asian Development Bank, *Managing Asian Cities* June 2008

3. Al Gore, *The Future. Six Drivers of Global Change*, Random House, January 2013

4. Hoornweg, Daniel and Bhada-Tata, Perinaz, World Bank, *What a Waste: A Global Review of Solid Waste Management*, March 2012

5. ibid

6. World Bank. Global Economic Prospects: *Managing the Next Wave of Globalization*, 2007

7. World Bank, *Global Economic Prospects*, 2009, Washington DC

8. World Bank, *World Development Report 2011: Resources Scarcity, Climate Change, and the Risk of Violent Conflict*, 2011

9. Global Water Security, Intelligence Community Assessment, February 2, 2012

10. World Bank, Global Economic Prospects: *Managing the Next Wave of Globalizaiton*, 2007

11. The *Guardian*, *What does the Arab World do when its water runs out?* February 19, 2011

12. World Bank, *Addressing China's Water Scarcity: Recommendations for Selected Water Resource Management Issue*, 2009

13. United Nations, *People and Lakes: Human Influences on Africa's Lakes*, 2006

14. The *New York Times*, *Water for Peace*, July 13, 2009

15. World Bank, *India's Water Economy: Bracing for a Turbulent Future*, 2006

16. World Bank, *Consumer Cooperatives: An Alternative Institutional Model for Delivery of Urban Water Supply and Sanitation Services*, January 2006

17. World Health Organization, *Safer Water Better Health: Costs, Benefits, and Sustainability of Interventions to Protect and Promote Health*, 2008

Running Out of Time

1. http://news.bbc.co.uk/2/hi/3575994.stm (Not on p. 20)

2. http://news.bbc.co.uk/2/southasia/6911544.stm(BasuKaushik)

3. http://www.igidr.ac.in/pdf/publication/WP-2011-009.pdf

4. http://www.imf.org/external/pubs/ft/reo/2012/APD/eng/areo0412.pdf

5. Evapo-transpiration means 'evaporation from earth surface to atmosphere from trees, plantation and agriculture'

6. http://www.albrightstonebridge.com/india-census/

7. The Census of India 2011 - Albright Stonebridge Group

8. Sacchidananda Mukherjee, Zankhana Shah, M. Dinesh Kumar, *Sustaining Urban Water Supplies in India: Increasing Role of Large Reservoirs.*

9. http://www.in.undp.org/content/dam/india/docs/india_urban_poverty_report_2009.pdf

10. MoHUPA- GoI and UNDP, (2009), *India-Urban Poverty Report 2009 Factsheet* (p. 7), New Delhi, Ministry of Housing and Urban Poverty Alleviation, United Nations Development Programme, Retrieved from http://www.in.undp.org/content/dam/india/docs/india_urban_poverty_report_2009.pdf

11. ibid

12. World Bank, 2011, *Towards lessons from the field in India drinking : lessons from the field* (p. 124), Washington D.C, The World Bank. Retrieved from http://documents.worldbank.org/curated/en/2011/01/16587753/towards-lessons-field-india-drinking-lessons-field

13. James, A. J. 2011, *India: Lessons for Rural Water Supply; assessing progress towards sustainable service delivery*, (p. 92), The Hague, Retrieved from http://www.waterservicesthatlast.org/content/download/374/2290/file/India country study.pdf

14. World Bank, 1999, *Rural Water Supply and Sanitation, Nature* (vol. 52)

15. Washington D.C, The World Bank, Retrieved from http://documents.worldbank.org/curated/en/1999/01/9334754/rural-water-supply-

16. Planning Commission, 2002, *India: Water Supply and Sanitation*, Government of India, New Delhi

17. Op.cit. James, A. J., 2011 p. 8

18. Ramanayya, T., Nagadevara, V., and Roy, S., 2008, *Study Related to Gap between the Irrigation Potential Created and Utilized* (p. 238), Bangalore

19. MoWR and GOI, 2006, *Sub-Committee more crop and income per drop of water* (p. 61). New Delhi, Ministry of Water Resources, Government of India

20. IDFC, 2011, *India Infrastructure Report 2011 Water: Policy and Performance for Sustainable development*, (p. 237), New Delhi, Oxford University Press. Retrieved from http://www.idfc.com/pdf/report/iir-2011.pdf

21. India Water Partnership, 2012, *Water, Agriculture and Climate Change, Consultation Meet on 'Water, Agriculture and Climate Change'* (p. 9), New Delhi, Institute of Human Development. Retrieved from http://cwp-india.org/pdf/Report on consultation meet on Water, Agriculture and Climate Change under APAN.pdf

22. op.cit., Ramanayya, T., Nagadevara, V., and Roy, S., 2008 p. 10

23. Planning Commission, 2006, *Report of the Steering Committee*

on Urban Development: For Eleventh Five Year Plan, 2007-12, p. 91, New Delhi

24. Perveen, S., Sen, R., and Ghosh, M. 2012, *Water Crisis : Water Risks for Indian Industries,* p. 19, Columbia Water Centre and FICCI. Retrieved from http://water.columbia.edu/files/2012/06/FICCI_CWC_IndiaWaterCrisisPaper.pdf

25. ibid

26. FICCI, 2011, *Water Use in Indian Industry Survey,* p. 7, New Delhi

27. op. cit. Perveen, S., Sen, R., and Ghosh, M., 2012 p. 12

28. ibid

29. ibid

30. ibid

31. Draft National Water Policy, 2012, recommended by National Water Board in its 14th Meeting, pp. 1–11, New Delhi, Ministry of Water Resource, Government of India. Retrieved from http://wrmin.nic.in/writereaddata/linkimages/DraftNWP2012_English9353289094.pdf

Growing Nation, Expanding Footprint

1. Stern, N., 2006, *The Economics of Climate Change: The Stern Review,* Cambridge, Cambridge University Press

2. Arnell, N. W., 2004, *Climate Change and Global Water Resources: SRES Emissions and Socio-Economic Scenarios,* Global Environmental Change, p. 14, pp. 31–52

3. Dennison, C., 2003, *From Beijing to Kyoto: Gendering the International Climate Change Negotiation Process,* 53rd Pugwash Conference on Science and World Affairs, *Advancing Human Security: The Role of Technology and Politics,* Canada, Halifax and Pugwash, Nova Scotia

4. op. cit. Hoekstra, A. Y., and A. K. Chapagain, 2008 p. 5

5. Hoekstra, A. Y. and Mekonnen, M. M., 2012, *The Water Footprint of Humanity,* (Peter H. Gleick, ed.), Proceedings of the

National Academy of Sciences of the United States of America, 109(9), 3232–7. doi:10.1073/pnas.1109936109

6. http://www.worldometers.info/water/

7. Approximation from the FAO, 2003, statistics to 2012, FAO, 2003, Agriculture Food and Water, p. 64, Washington D.C, Food and Agriculture Organisation. Retrieved from ftp://ftp.fao.org/agl/aglw/docs/agricfoodwater.pdf

8. ibid

9. ibid

10. 2030 Water Resource Group, 2009, *Charting our water future : economic frameworks to inform decision-making*, pp. vii, 185, Electronic resource: McKinsey & Company. Retrieved from http://www.mckinsey.com/App_Media/Reports/Water/Charting_Our_Water_Future_Full_Report_001.pdf

11. ibid

12. ibid

13. ibid

14. ibid

15. ibid

16. Amarasinghe, U. A., Shah, T., Turral, H., and Anand, B. K., 2007, *India's water future to 2025-2050: business-as-usual scenario and deviations*, p. 41, Colombo, Sri Lanka, International Water Management Institute (IWMI Research Report 123). doi:10.3910/2009.124

17. ibid

18. op. cit. Amarasinghe, *et.al*, 2007

19. op. cit. 2030 Water Resource Group, 2009

20. This section is sourced from *Mahindra Rise*, 2011, *Alternative Accelerates Results: Sustainability Review 2011 – 12,* Mumbai, Mahindra & Mahindra Limited

21. Mahindra & Mahindra Limited - Automotive Division (AD)

22. Systems & Technologies Sector

23. Mahindra Lifespace Developers Limited

24. Financial Services Sector

25. Mahindra Vehicle Manufacturers Limited

26. *Mahindra Rise*, 2011, *Alternative Accelerates Results: Sustainability Review 2011–12*, Mahindra & Mahindra Limited

27. Mahindra Group

28. *Mahindra Rise*, 2011, *Alternative Accelerates Results: Sustainability Review 2011–12*, Mahindra & Mahindra Limited

29. Indian Hotels Company Limited, 2011, *Beyond The Numbers; 8th Corporate Sustainability Report*, Indian Hotels Company Limited

30. ibid

31. This section is sourced from HCC, 2011, *Responsible Infrastructure: HCC Sustainability Report 2010-11,* Hindustan Construction Co. Ltd

32. HCC, 2011, *Responsible Infrastructure: HCC Sustainability Report 2010-11*, Hindustan Construction Co. Ltd

33. Tata Chemicals, 2011, *Tata Chemicals: Sustainability Report 2088-2010*, Tata Chemicals

The Pool is getting murkier

1. Central Pollution Control Board, 2009, *Status of Water Quality in India- 2009* Delhi, Central Pollution Control Board, Ministry of Environment and Forestry. Retrieved from http://www.cpcb.nic.in/upload/NewItems/NewItem_169_waterquality.pdf

2. *TOI* September 24, 2013, *One-fifth of centrally-funded sewage treatment plants remain non-operational*

3. UN Water, 2009, *Water in a Changing World* (3rd edition), Washington D.C, The United Nations World Water Development Report

4. UNESCO–WWAP, 2003, *Water for People Water for Life*, Washington D.C, The United Nations World Water Development Report, Kumar, S., and Murty, M. N., 2011, *Water Pollution in India: An Economic Appraisal, India Infrastructure Report 2011, Water: Policy and Performance for Sustainable Development.* Oxford University Press

and IDFC. Retrieved from http://www.idfc.com/pdf/report/2011/Chp-19-Water-Pollution-in-India-An-Economic-Appraisal.pdf

5. ibid
6. op.cit., p. 9, UN Water, 2009
7. op.cit., p. 8, UN Water, 2006
8. op.cit., p. 10, Kumar & Murty, 2011
9. op.cit., p. 8, UN Water, 2006
10. ibid
11. ibid
12. ibid
13. ibid
14. Central Pollution Control Board, *Status of Water Quality in India-2009*, Ministry of Environment and Forestry, Delhi
15. Central Pollution Control Board, 2009, *Status of Water Quality in India- 2009*, Ministry of Environment and Forestry. Retrieved from http://www.cpcb.nic.in/upload/NewItems/NewItem_169_waterquality.pdf
16. op.cit., UN Water, 2006

The Policy Tangle

1. *Draft National Policy*, as recommended by the National Water Board, June, 2012
2. G N Kathpalia, Rakesh Kapoor, *Water Policy and Action Plan for India 2020: An Alternative*, November, 2002
3. Shiney Varghese, *Corporatizing Water: India's Draft National Water Policy*, February 2012
4. Phil Mader, *Water Paradigms: Full Cost Recovery versus Human Rights*
5. China Water Risk web portal
6. *Good Practices in Urban Water Management*, Asian Development Bank Study, 2012
7. Singapore's National Water Agency, PUB's portal

8. *Recognizing the Strategic Value of Water*, CDP Water Programme South Africa Report, 2012

Pricing the Elixir

1. *An Integrated Assessment of Water Markets: Australia, Chile, China, South Africa and the USA*, R. Quentin Grafton, Clay Landry, Gary Libecap, Sam McGlennon, Bob O'Brien, 2010

2. *Water and Wastewater Pricing*, Holly Stallworth, Economist, Office of Wastewater Management, 2009

3. *Draft National Policy*, as recommended by the National Water Board, June, 2012

4. *Analysis & Reviews*, China Water Risk portal

5. *Water Pricing towards Sustainability of Water Resources: A Case Study in Beijing*, Shao Liu, 2000

6. *Administered Prices-Water*, Rolfe Eberhard, for National Treasury South Africa

7. Singapore's National Water Agency, PUB's portal

8. *Review of pricing reform in the Australian water sector*, National Water Commission, Australia, 2011

9. *Are the Prices Right? Balancing Efficiency, Equity, and Sustainability in Water Pricing*, Steven Renzetti, 2007

10. *Water Pricing as a Demand Management Option: Potentials, Problems and Prospects*, V. Ratna Reddy, Centre for Economic and Social Studies, Hyderabad

11. *Pricing water resources to finance their sustainable management*, EU Water Initiative–Finance Working Group, May, 2012

12. Water, US EPA Portal

13. *International Water Pricing: An Overview and Historic and Modern Case Studies*, Kristin M. Anderson1 and Lisa J. Gaines, 2009

14. *Managing Water for All- An OECD Perspective on Pricing and Financing*, 2009

Virtual Web of Water Trade

1. UN Water, 2009

2. Allan, J.A., 1993, *Priorities for Water Resources Allocation and Management*, Overseas Development Administration (ODA), London, Allan, J.A., 1994, *Overall perspectives on countries and regions*, Rogers, P., Lydon, P. (eds.), *Water in the Arab World: Perspectives and prognoses*, Harvard University Press, Cambridge, MA

3. Hoekstra, A. Y., and Chapagain, A. K., 2006, *Water footprints of nations: Water use by people as a function of their consumption pattern*

4. UN Water, 2009, *Water in a Changing World* (3rd edition), Washington D.C, The United Nations World Water Development Report

5. Verma, S., Kampman, D. A., Van der Zaag, P., and Hoekstra, A. Y., 2009, *Going against the flow: A critical analysis of inter-state virtual water trade in the context of India's National River Linking Program. Physics and Chemistry of the Earth*, Parts A/B/C

6. ibid

7. ibid

8. ibid

9. De Fraiture CCAi, Amarasinghe, U rosegrant, Molden D

10. op.cit., Verma, *et al.*, 2009

11. op.cit., Hoekstra and Chapagain, 2006

12. op.cit., Verma, *et al.*, 2009

13. http://www.waterfootprint.org/?page=files/India

14. http://www.waterfootprint.org/?page=files/India

15. Zimmer, D., and Renault, D., 2003, *Virtual water in food production and global trade: Review of methodological issues and preliminary results*. In Hoekstra, A.Y. (ed.), *Virtual water trade: Proceedings of the International Expert Meeting on Virtual Water Trade*, Value of Water Research Report Series no. 12, UNESCO-IHE, Delft

16. ibid

17. op.cit., Hoekstra, Arjen Y, and Mekonnen, M. M., 2012

18. Yang, H., Wang, L., Abbaspour, K. C., and Zehnder, A. J. B., 2006, *Virtual water trade: an assessment of water use efficiency in the international food trade*, Hydrology and Earth System Sciences

19. ibid

20. ibid

21. Allan, J.A., 2003, *Virtual Water-the Water, Food, and Trade Nexus, Useful Concept or Misleading Metaphor?*, Water International

22. ibid

Bottle With a Narrow Neck

1. *Water Infrastructure: Enhancing Regional Cooperation in Infrastructure Development including that Related to Disaster Management*, UNESCAP

2. *The Water Poverty Index: an International Comparison,* Peter Lawrence, Jeremy Meigh and Caroline Sullivan, Keele Economics Research Papers, 2002

3. *Irrigation Impacts on Income Inequality and Poverty Alleviation*, Madhusudan Bhattarai, R. Sakthivadivel and Intizar Hussain, 2002

4. *India Water Supply and Sanitation*, Energy and Infrastructure Department, World Bank, 2006

5. *Water: Policy and Performance for Sustainable Development*, India Infrastructure Report, by IDFC, 2011

6. *Sustaining India's Irrigation Infrastructure*, National Centre for Agricultural Economics and Policy Research

7. Centre for Science and Environment Web Portal

8. *Water Sector in India*, Ernst & Young Report, 2011

9. *Water Infrastructure*-US EPA web portal

10. *Foreign Investment in China's Water Infrastructure: A Strategy for National Security*, Yusha Hu, 2010

11. PAC-Growth Acceleration Programme, Government of Brazil

12. *Private–Public Partnerships in Urban Water Supply Sector: A Study of the Regional Trends*, Asanga Gunawansa, Sonia Ferdous Hoque, Lovleen Bhullar, 2012

Acknowledgements

We are grateful to Indian Merchants Chamber (IMC) including its research wing, Economic Research & Training Foundation; and Responsible Investment Research Association (RIRA) for providing a wonderful opportunity to initiate the research that eventually developed into this book.

There are a number of water experts including those with ground level experience, with whom the authors have been sharing views on different aspects related to water, and have been greatly benefitted by continued interactions with them, spread across a period of over a year. In this regard, the authors wish to express sincere thanks particularly to Dr Abhay Pethe (renowned economist, Mumbai University), Mr Rajeev Jalota, Additional Municipal Commissioner MCGM-Projects, Water Supply, Sewerage and Sanitation; Dr D T Dange, Principal Advisor-Water Supply Resource Management Cell-MMRDA; Dr Rakesh Kumar, Chief Scientist and Head, NEERI Mumbai; Mr Nishith Desai, Head, Nishith Desai Associates; Mr Shawahiq Siddiqui, Lawyer Environment; Mr Ramani Iyer, Mentor for Water with Forbes Campbell); Dr Ram Babu, CEO, Carbon Advisory, Mumbai and Mr D G Sonwane, Tata Consulting Engineers.

We are also indebted to Mr Mangesh Gupte, Sr General Manager, Head, CSR and Water Sustainability, ACC limited;

Dr R Mopalwaar, Director MJP; Mr Sameer Unhale, JNNURM, MMRDA and Mr Shirish Garud, TERI, Mumbai for guiding us in the right path for this important research for the inclusive and sustainable growth of Indian economy.

Our special thanks are also due to Mr Sushil Torne, (PhD student at Mumbai University), Mr Rohan Oak and Ms Vaishali Gautam (Programme Assistant at RIRA) for their sincere efforts in providing consultation on case stories and background research.

A Note on the Authors

Joseph P. Quinlan is Managing Director and Chief Market Strategist at US Trust, Bank of America, Private Wealth Management. In this role, Joe oversees the development and implementation of macro-investment strategies, with a particular focus on US sector strategies, the emerging markets, and developed markets. He is also responsible for the firm's global thematic research. His research centres on both the US and the global economy, and is frequently cited in such media venues as *Barron's*, *The Wall Street Journal*, and the *Financial Times*.

With nearly 30 years of financial services experience, Mr Quinlan, most recently served as a senior global economist/strategist for Morgan Stanley. Mr Quinlan started his career with Merrill Lynch.

Mr Quinlan lectures on finance and global economics at the New York University, where he has been a faculty member since 1992. He is also a professor at Fordham University and American University in Washington DC. In 2012, he was appointed a Visiting Professor at Wuhan University, Wuhan, China. He regularly lectures at various universities around the world.

In 1998, he was nominated as an Eisenhower Fellow. Presently, he is a Senior Fellow at the Paul H. Nitze School of Advanced International Studies of Johns Hopkins University, Washington D.C., and a Senior Fellow at the German Marshal Fund in Brussels, Belgium. He was appointed a Bosch Fellow at the Transatlantic Academy in December 2001.

In 2006, the American Chamber of Commerce to the European Union awarded Mr Quinlan, the 2006 Transatlantic Business Award, for his research on US-Europe economic ties. In 2007, he was a recipient of the European-American Business Council Leadership award for his research on the transatlantic partnership and global economy.

Mr Quinlan regularly debriefs policy makers and legislators on Capitol Hill on global trade and economic issues. He has testified before the European Parliament. He has served as a consultant to the US Department of State, and served as the US representative (Economic Policy Committee) to the Organization for Economic Cooperation and Development in Paris, France, for the US Council for International Business. He is also a board member of Fordham University's Graduate School of Arts and Science and serves on Fordham University's President Council.

He is the author, co-author, or contributor to twenty books. His most recent book, *The Last Economic Superpower,* was released by McGraw Hill, in November 2010. He has published more than 125 articles on economics, trade, and finance that have appeared in such venues as *Foreign Affairs*, the *Financial Times*, *The Wall Street Journal* and *Barron's*. He regularly appears on *CNBC*, as well as *Bloomberg* television, *PBS* and other media venues.

He has earned his B.A. in Political Science/International Affairs from Niagara University and M.A. in International Political Economics and Development from Fordham University.

Sumantra Sen is the Founder and Chief Executive Officer at Responsible Investment Research Association (RIRA), a non-profit investor forum that facilitates mainstreaming of Environment, Social and Governance (ESG) factors into financial markets in India, through active inputs in policy, research and capacity building. He also serves as Director at Penn Schoen Berland, a research-based, strategic communication firm, where he leads the Financial Communication and Capital Markets Advisory practice.

Mr Sen is an MBA and a member of the Chartered Institute of Securities and Management, UK. In his career of over 20 years with leading financial organizations like ABN AMRO and Bank of America, he has wide-ranging experience in global capital markets and investment management.

Mr Sen has subject matter interest and expertise in corporate governance and responsible investment. He serves on the advisory board at Association for Sustainable and Responsible Investment in Asia (ASrIA), Hong Kong, and is an advisor to the Climate Bond Initiative, UK. He is a member of the Academic Network at United Nations Principles for Responsible Investments (UNPRI). Recently, he has taken a honourary leadership role at LiberalsIndia for Good Governance, a policy think-tank.

Mr Sen contributes to published work in the *Responsible Investment and Sustainable Development* regularly. In addition to a chapter on India's ESG issues in *Wiley Finance's, Evolutions in Sustainable Investing*, he is involved in several upcoming book projects and is associated with international publishers. He is also working closely with leading television broadcast companies to develop content around sustainable development, climate change and inclusive growth.

Mr Sen's expert comments and articles have been covered in *The Economic Times*, *SRI-Connect*, *Ethical Markets*, *Clean Biz Asia*, and *Responsible Investor and Benchmark* amongst other media resources. He is also a frequent speaker on these themes at domestic and international investment summits.

Kiran Nanda is a corporate economic and sustainability consultant, who till recently has been Director at Indian Merchant's Chamber-Economic Research and Training Foundation (IMC-ERTF) for about 8 years. At IMC-ERTF, she handled the IMC's Energy and Environment Committee, which comprises of distinguished experts and professionals in environment, water, and energy domain. Also was particularly responsible for water economics applied research. She has conducted a number of seminars on water, energy among other sustainability national issues. Before joining IMC, she was Chairperson-Economic and Political Affairs Committee of IMC for about a decade.

Ms Nanda has over 30 years of wide range of experience in applied economics. She has worked with Ambuja Cement Ltd for almost 13 years as Chief Economist heading the Corporate Economics Division. She has represented cement industry's key concerns on housing and infrastructure to the concerned authorities, both at the centre and the western region, and has been in close touch with various leading Chambers of Commerce and Industry.

She served as Senior Economist in Tata's Economics and Statistics Department for over 16 years. Earlier, she worked as a researcher in the Economics Research Bureau of *The Economic Times* for a period of 3 years. She has also worked as the economist member of a high powered regional assessment panel on appraisal of Indian projects of DFID, the UK government's Financial Deepening Challenge Fund.

Ms Nanda's research focuses on applied economic research in corporate sustainability subjects, including policy research with respect to micro and macro linkages; agri-business management; CSR, water, renewables, infrastructure development, SME sector etc. Having worked all along in the corporate sector and government bodies, she has contributed to the evolution of the framework and policies on select burning issues in the field of cement, housing and infrastructure.

Her work has been extensively recognized and acknowledged. She headed and made a significant contribution to several research projects such as, 'Towards Sustainable Practices and Strategies for Efficient Water Resources Management for Mumbai', 'Strategies and Preparedness for Trade and Globalization in India', 'Supply Chain Management for Indian Agriculture, 'Corporate Initiatives in Rural and Agricultural Transformation and What Indian SMEs Need?'. At present, she advises corporate houses and contributes to corporate sustainability publications.